T0297424

Studies in Systems, Decision and Control

Volume 47

Series editor

Janusz Kacprzyk, Polish Academy of Sciences, Warsaw, Poland
e-mail: kacprzyk@ibspan.waw.pl

About this Series

The series "Studies in Systems, Decision and Control" (SSDC) covers both new developments and advances, as well as the state of the art, in the various areas of broadly perceived systems, decision making and control- quickly, up to date and with a high quality. The intent is to cover the theory, applications, and perspectives on the state of the art and future developments relevant to systems, decision making, control, complex processes and related areas, as embedded in the fields of engineering, computer science, physics, economics, social and life sciences, as well as the paradigms and methodologies behind them. The series contains monographs, textbooks, lecture notes and edited volumes in systems, decision making and control spanning the areas of Cyber-Physical Systems, Autonomous Systems, Sensor Networks, Control Systems, Energy Systems, Automotive Systems, Biological Systems, Vehicular Networking and Connected Vehicles, Aerospace Systems, Automation, Manufacturing, Smart Grids, Nonlinear Systems, Power Systems, Robotics, Social Systems, Economic Systems and other. Of particular value to both the contributors and the readership are the short publication timeframe and the world-wide distribution and exposure which enable both a wide and rapid dissemination of research output.

More information about this series at http://www.springer.com/series/13304

D.A. Novikov

Cybernetics

From Past to Future

 Springer

D.A. Novikov
Trapeznikov Institute of Control Sciences
 RAS
Moscow
Russia

ISSN 2198-4182   ISSN 2198-4190   (electronic)
Studies in Systems, Decision and Control
ISBN 978-3-319-27396-9   ISBN 978-3-319-27397-6   (eBook)
DOI 10.1007/978-3-319-27397-6

Library of Congress Control Number: 2015957121

Printed on acid-free paper

This Springer imprint is published by SpringerNature
The registered company is Springer International Publishing AG Switzerland

*In warm memory of my father,
Academician A.M. Novikov, who opened up
the world of cybernetics to me*

Contents

About the Author

D.A. Novikov Dr. Sci. (Eng.), Prof., born in 1970. A corresponding member of the Russian Academy of Sciences, deputy director of Trapeznikov Institute of Control Sciences of the Russian Academy of Sciences, and head of Control Sciences Department at Moscow Institute of Physics and Technology.

Author of over 500 scientific publications on theory of control for interdisciplinary systems, methodology, systems analysis, game theory, decision-making, project management, and control mechanisms for organizational and socioeconomic systems.
e-mail: novikov@ipu.ru, www.mtas.ru

Introduction

We had dreamed for years of an institution of independent scientists, working together in one of these backwoods of science, not as subordinates of some great executive officer, but joined by the desire, indeed by the spiritual necessity, to understand the region as a whole, and to lend one another the strength of that understanding. N. Wiener

The history of science development has *"romantic" periods*. One of them fell on the middle of the 1940s. "Romanticism" was determined by several factors.

The first factor concerned *an intensive flow of scientific and applied results*. Just imagine: the end of the terrible World War II (1945); dynamic growth of industry; the way out of the crisis in physics (which occurred at the beginning of the twentieth century)—the appearance and rapid development of atomic physics, quantum mechanics, general and special relativity theory, and astrophysics; first atomic bomb explosion (1945), followed by first atomic power plant launch (1954); electrical and radio devices usage by everymen; a series of discoveries in biology, physiology, and medicine (commercially produced (1941) penicillin (1928) saved millions of lives, the soon appearance of the three-dimensional DNA helix model (1953), rapid development of radiobiology and genetics, etc.); creation of first computer (1945) and bipolar transistor (1947); and the birth of choice theory [12] (1951), artificial neural networks (1943), game theory (1944–see [143, 145]), and operations research (1943), representing a striking example of an interdisciplinary synthetic science.

The second factor was associated with the comprehension of science *interdisciplinarity* by researchers from different branches. Interdisciplinarity implies that (a) there exist general approaches and regularities in different scientific branches and (b) it is possible to perform an *adapted translation of results* between some branches. This led to the obvious necessity of generalization search, not only within the framework of a certain field of knowledge or at a junction of fields, but (in the first place) at their "intersection." In other words, the matter was not even to create new paradigms in T. Kuhn's sense [112] for a branch, but to apply joint efforts of physicians and biologists, mathematicians, engineers and physiologists, etc., for obtaining fundamentally new results and breakthrough technologies.

The third factor was that the *role and "benefit" of science became evident for everymen and politicians*. The former enjoyed scientific results owing to their rapid and mass implementation. The latter (a) realized that science is an important public and economic drive of a society and (b) got accustomed to that project-based management of applied research allows predicting and in part guaranteeing its duration and results.

However, the flight of thought and stormy feelings of any romanticism go in parallel with *overestimated expectations*. Moreover, the onrush development of any science is inevitably followed by its normal advancement (e.g., according to T. Kuhn).

All these regularities fully affected *cybernetics*—a science born in the above "romantic period" (1948) and undergone romantic childhood, the disillusionment of juvenility and the decay of maturity.[1] The book discusses exactly these issues, representing a brief "navigator" across the history of cybernetics, its state of the art, and prospects. The style of a "navigator" implies renunciation of a detailed characterization of results: Numerous references cover almost all[2] classical works on cybernetics[3] published to date. On the other hand, such style a priori makes exposition somewhat incomplete, eclectic, and nonrigorous, as it would seem to a representative of any concrete science mentioned.

The book possesses the following structure. First, we consider the evolution of cybernetics (from N. Wiener to the present day), see Sects. 1.1 and 1.2. A detailed analysis focuses on the reasons of its ups and downs in Sect. 1.3. Next, we study the interrelation of cybernetics with *control philosophy* and *control methodology* (Chap. 2), as well as with *systems theory* and *systems analysis* (Chap. 4). Chapter 3 discusses the basic laws, regularities, and principles of control. Chapter 5 identifies some modern *development trends* of cybernetics. In the conclusion, we introduce the new stage of cybernetics development—the so-called *Cybernetics 2.0* as the science of systems organization and control. Appendices contain a list of basic terms and topics for self-study.

The author is deeply grateful to V. Afanas'ev, V. Breer, V. Burkov, A. Chkhartishvili, M. Goubko, A. Kalashnikov, K. Kolin, V. Kondrat'ev, N. Korgin, O. Kuznetsov, A. Makarenko, R. Nizhegorodtsev, B. Polyak, I. Pospelov, A. Raikov, P. Skobelev, A. Teslinov, and V. Vittikh for fruitful discussions and valuable remarks. Of course, the author takes all shortcomings as referring to himself.

And finally, my deep appreciation belongs to A. Mazurov for his careful translation, permanent feedback, and contribution to the English version of the book.

---

[1]Note that general systems theory and systems analysis proceeded a similar path, see below.

[2]Cybernetics is a synthetic science and any attempt to give a comprehensive bibliography of its components (e.g., control theory) is doomed to failure. By saying "all," we mean cybernetics proper (Cybernetics with capital C as explained in Sect. 1.1).

[3]Most references are publicly available to an interested reader in Internet.

Chapter 1
Cybernetics in the 20th Century

This section is intended to consider in brief the history of cybernetics and describe "classical" cybernetics. Let us call it "*cybernetics 1.0*".

CYBERNETICS (from the Greek κυβερνητική "governance," κυβερνώ "to steer, navigate or govern," κυβερνη "an administrative unit; an object of governance containing people"[1]) is the science of general regularities of *control* and *information transmission* processes in different systems, whether *machines, animals* or *society*.

Cybernetics studies the concepts of control and communication in living organisms, machines and organizations including self-organization. It focuses on how a (digital, mechanical or biological) system processes information, responds to it and changes or being changed for better functioning (including control and communication).

Cybernetics is an *interdisciplinary science*. It originated "at the junction"[2] of mathematics, logic, semiotics, physiology, biology and sociology. Among its inherent features, we mention analysis and revelation of general principles and approaches in scientific cognition. Control theory, communication theory, operations research and others (see Sect. 1.1) represent most weighty theories within cybernetics 1.0.

In ancient Greece, the term "cybernetics" denoted the art of a municipal governor (e.g., in Plato's *Laws*).

A. Ampere (1834) related cybernetics to political sciences: the book [6] defined cybernetics ("the science of civil government") as a science of current policy and practical governance in a state or society.
B. Trentowsky (1843, see [136, 201]) viewed cybernetics as "the art of how to govern a nation."

In its *Tektology* (1925, see [29]), A. Bogdanov examined common organizational principles for all types of systems. In fact, he anticipated many results of N. Wiener and L. Bertalanffy, as the both were not familiar with Bogdanov's works.

---

[1]This root induced the words "governor", "government" and "governance."
[2]Depending on the mutual penetration of subjects and methods, a pair of sciences often appears at the junction of two sciences (e.g., physical chemistry and chemical physics).

© Springer International Publishing Switzerland 2016
D.A. Novikov, *Cybernetics*, Studies in Systems, Decision and Control 47,
DOI 10.1007/978-3-319-27397-6_1

The *history* and evolution of *cybernetics* can be traced in [65, 84, 85, 168, 179, 206].

The modern (and classical!) interpretation of the term "cybernetics" as "the scientific study of control and communication in the animal and the machine" was pioneered by *Norbert Wiener* in 1948, see the monograph [221]. Two years later, Wiener also added society as the third object of cybernetics [225]. Among other classics, we mention *William Ashby*[3] [14, 15] (1956) and *Stafford Beer* [23] (1959), who made their emphasis on the biological and "economic" aspects of cybernetics, respectively.

Therefore, cybernetics 1.0 (or simply **cybernetics**) can be defined[4] as "THE SCIENCE OF CONTROL AND DATA PROCESSING IN ANIMALS, MACHINES AND SOCIETY." An alternative is the definition of **Cybernetics** (with capital C, to distinguish it from cybernetics whenever confusion may occur) as "THE SCIENCE OF *GENERAL REGULARITIES* OF CONTROL AND DATA PROCESSING IN ANIMALS, MACHINES AND SOCIETY." The second definition differs from its first counterpart in the words "general regularities," which is crucial and will be repeatedly underlined and used below. In the former case, the matter concerns "the umbrella brand," i.e., the "integrated" results of all sciences dealing with problems of control and data processing in animals, machines and society. The latter case covers partial "intersection" of these results[5] (see Fig. 1.9), i.e., usage of common results for all component sciences. Furthermore, we will adhere to this approach over and over again for discrimination between the corresponding umbrella brand and the common results of all component sciences in the context of different categories such as interdisciplinarity, systems analysis, organization theory, etc.

1.1 Wiener's Cybernetics

Some historical facts (an epistemological view). Any science is determined by its "*subject*" (problem domain) and "*method*" (an integrated set of methods) [112, 131, 148]. Therefore, sciences[6] can be divided into:

---

[3]Interestingly, W. Ashby introduced and analyzed the categories of "variety" and "self-organization," as well as the terms "homeostat" and "black box" in cybernetics.

[4]These definitions will be addressed throughout the whole book, except the Conclusion.

[5]Figuratively speaking, the central rode of the "umbrella."

[6]This somewhat conditional differentiation applies not only to sciences, but to researchers. As mentioned in [149], in some fields of science researchers are traditionally divided into two categories. The first one is called "screwmen." They study new problem domains ("screws") using common methods ("spanners"). The second category is known as "spannermen"; such researchers design new technologies of cognition (methods, "spanners") and illustrate their efficiency in different problem domains (for unscrewing common "screws").

- *subject-oriented sciences* studying a certain subject by different methods (e.g., physics, biology, sociology);
- *method-oriented sciences* (in the restricted sense, the so-called *model-based sciences*) developing a certain set of methods applicable to different subjects; for instance, a classical example is applied mathematics: the apparatus and methods of its branches (differential equations, game theory, etc.) serve for description and analysis of different-nature systems;
- *synthetic sciences* ("*metasciences*") mostly developing and/or generalizing methods of certain sciences in application to subjects of these and/or other sciences (e.g., operations research, systems analysis, cybernetics). With the course of time, synthetic sciences find or generate their "own" subjects and methods.

As sciences of any types develop, their subjects and methods are split and intersected by each other, causing further differentiation of sciences.

The following conditions guarantee the appearance (1 and 2) and survival (3) of synthetic sciences:

1. A sufficient development level of origin/*source sciences*;
2. Numerous *analogies* (and then *generalizations*) among partial results of source sciences;
3. Rather easy and fast generation/accumulation of nontrivial theoretical and applied results and their popularization, within the scientific community and everymen.

Speaking about cybernetics, the first and second conditions had been satisfied by the middle of the 1940s (see the Introduction). And the long-term cooperation between N. Wiener and biologists, alongside with his wide and deep professional interests (recall Wiener processes, Banach-Wiener spaces, the Wiener-Hopf equations) ensured "subjective" satisfaction of these conditions. In its late interview to *Russian Studies in Philosophy* (1960, No. 9), Wiener noted that "the aim was to unite efforts in different branches of science and get focused on *uniform solution of similar problems.*" The third condition–rapid accumulation and popularization of new results–was also realized, see the discussion below.

In 1948 integration of results obtained by different sciences and their substantiated applicability to different subjects (see Fig. 1.1) gave birth to a new synthetic science known as *Wiener's cybernetics*.

A science as a system of knowledge has the following *epistemological functions* [149]:

- *descriptive* (phenomenological) function, i.e., acquisition and accumulation of data and facts. Any science starts from this function, viz., answering to the question "What is the structure of the world?", as any science can be based on

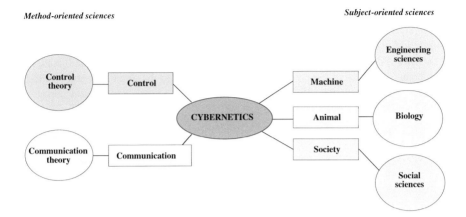

Fig. 1.1 The phylogenesis of Wiener's cybernetics

very many facts. From this viewpoint, cybernetics as a synthetic science[7] mostly employs the results of its components (source sciences);

- *explanatory* (explicative) function, i.e., elucidation of phenomena and processes, their internal mechanisms. Here the question to-be-answered is "Why does the world is exactly this?" In this function, cybernetics plays a more visible role: even analogies may have powerful elucidation;
- *generalizing* function, i.e., formulation of laws and regularities systematizing and absorbing numerous fragmented phenomena and facts (the associated question is "What are the common features of… ?"). Perhaps, this is the main function of cybernetics, since generalizations (in the form of laws, regularities, models, research approaches) comprise the framework of its results;
- *predictive* (prognostic) function, i.e., scientific knowledge allow predicting new processes and phenomena (this function answers the question "What and why will happen?"). Efficient forecasting is possible using substantiated analogies and constructive generalizations within synthetic science cybernetics;
- *prescriptive* (normative) function, i.e., scientific knowledge allow organizing activity with certain goals (the corresponding question is "What and how should be done for goal achievement?"). Normative function has a close connection with solution of control problems, an important subject of cybernetics.

Definitions. Just like any comprehensive category, cybernetics would hardly possess a unique definition. Moreover, the meanings of terms describing this category also evolve with the course of time. Let us give a series of widespread definitions of cybernetics:

A science concerned with the study of systems of any nature which are capable of receiving, storing, and processing information so as to use it for control–A. Kolmogorov;

---

[7]For instance, A. Kolmogorov believed that cybernetics is not a science but a scientific direction; however, the listed functions also apply to the latter.

The art of steersmanship: deals with all forms of behavior in so far as they are regular, or determinate, or reproducible: stands to the real machine–electronic, mechanical, neural, or economic–much as geometry stands to real object in our terrestrial space; offers a method for the scientific treatment of the system in which complexity is outstanding and too important to be ignored.–W. Ashby;

A branch of mathematics dealing with problems of control, recursiveness, and information, focuses on forms and the patterns that connect.–G. Bateson;

The art of effective organization.–S. Beer;

The art of securing efficient operation.–L. Couffignal;

The art and science of manipulating defensible metaphors.–G. Pask;

The art of creating equilibrium in a world of constraints and possibilities.– E. Glasersfeld;

The science and art of understanding.–H. Maturana;

A synthetic science of control, information and systems–A.G. Butkovsky;

A system of views a governor must have for efficient control of its κυβερνη– N. Moiseev;

The art of interaction in dynamic networks.–R. Ascott.

Almost all definitions involve the terms "control" and "system," see the definition of "cybernetics 2.0" in the Conclusion. Therefore, they are mutually non-contradictory and well consistent with the definition of cybernetics accepted by us.

Consequently, Wiener's cybernetics has the following key terms: control, communication, system, information, feedback, black box, variety, homeostat.

Cybernetics today (disciplines included in cybernetics in the descending order of their "grades" of membership, see Fig. 1.9, with year of birth if available):

- *control theory*[8] (1868–the papers [127, 216] published by J. Maxwell and I. Vyshnegradsky);
- *mathematical theory of communication and information* (1948–Shannon's works [187, 188]);
- *general systems theory, systems engineering and systems analysis*[9] (1968–the book [26] and 1956–the book [92]);
- *optimization* (including linear and nonlinear programming; dynamic programming; optimal control; fuzzy optimization; discrete optimization, genetic algorithms, and so on);
- *operations research* (graph theory, game theory and statistical decisions, etc.);
- *artificial intelligence* (1956–The Dartmouth Summer Research Project on Artificial Intelligence);
- *data analysis* and *decision-making*;
- *robotics*

and others (purely mathematical and applied sciences and scientific directions, in an arbitrary order) including systems engineering, recognition, artificial neural networks and neural computers, ergatic systems, fuzzy systems (rough sets, grey

---

[8]According to an established tradition, control science will be called control theory (yet, such name narrows its subject).

[9]Chapter 4 discusses the history of these scientific directions in more details.

systems [91, 94, 162, 165]), mathematical logic, identification theory, algorithm theory, scheduling theory and queuing theory, mathematical linguistics, programming theory, synergetics and all that jazz.

In its components, cybernetics intersects considerably with many other sciences, in the first place, with such metasciences as general systems theory and systems analysis (see Chap. 4) and *informatics*[10] (see the Conclusion).

There exist a few classical monographs and textbooks on Cybernetics with its "own" results; here we refer to [2, 14, 22, 23, 26, 62, 63, 133, 222–225]. On the other hand, textbooks on cybernetics (mostly published in the former USSR) include many of the above-mentioned directions (*par excellence*, control in technical systems and informatics)–see [52, 68, 108, 113, 119].

The prefix *"cyber"* induces new terms on a regular basis, viz., cybersystem, cyberspace, cyberthreat, cybersecurity, etc. In a broader view of things, this prefix embraces all connected with automation, computers, virtual reality, Internet and so on.[11]

Nowadays, cybernetics attracts the attention of several hundreds of dedicated *research centers* (institutes, departments, research groups) and associations[12] worldwide (with explicit usage of the term "cybernetics" in their names), plus hundreds of *scientific journals* and regular *conferences*. For instance, see Internet resources on cybernetics:

- http://www.asc-cybernetics.org/
- http://pespmc1.vub.ac.be/
- http://wosc.co/
- http://neocybernetics.com/wp/links/

and others.

"Sectoral" types of cybernetics. Alongside with general cybernetics, there exist *special ("sectoral") types of cybernetics* [113]. A most natural approach (which follows from Wiener's extended definition) is to separate out technical cybernetics, biological cybernetics and socioeconomic cybernetics besides *theoretical cybernetics* (i.e., Cybernetics).

It is possible to compile a more complete list of special types of cybernetics (in the descending order of the current level of exploration):

- *technical cybernetics, engineering cybernetics*;
- *biological* and *medical cybernetics*, evolutionary cybernetics, cybernetics in psychology [5, 9, 10, 15, 24, 61, 100, 109, 160, 169, 202];

---

[10]Or even with computer science, but we will omit this aggregative term due to its undetermined and eclectic character.

[11]Perhaps, this reflects the word "cybernetics" in mass consciousness, even despite that experts in the field disagree with such (general and simplified) usage of the prefix.

[12]Principia Cybernetica (V. Turchin et al.), American Society for Cybernetics (http://www.asc-cybernetics.org), World Organization of Systems and Cybernetics, to name a few.

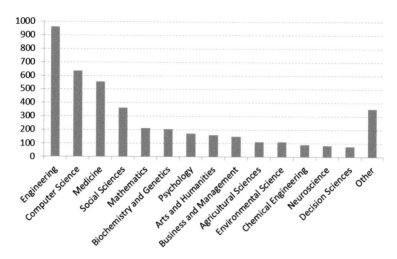

Fig. 1.2 The usage of the term "Cybernetics" by scientific branches in paper titles indexed by Scopus

- *economic cybernetics* [22, 23, 99, 138, 227];
- physical cybernetics (to be more precise, *"cybernetical physics"*,[13] see [203]);
- *social cybernetics, educational cybernetics*;
- *quantum cybernetics* (quantum systems control, quantum computing) (see surveys in [69, 72]).

As standing apart, we mention a branch of biological cybernetics known as *cybernetic brain modeling* integrated with artificial intelligence, neural and cognitive sciences. A romantic idea to create a cybernetic (computer-aided) brain at least partially resembling a natural brain stimulated the founding fathers of cybernetics (see the works of Ashby [15], Walter [218], Arbib [11], George [61], Steinbuch [193] and others) and their followers (for a modern overview, we refer to [169]).

Bibliometric analysis. The degree of penetration of cybernetics into other sciences and the scale of its "synthetic" character can be estimated using a simple bibliometric analysis. Figures 1.2 and 1.3 demonstrate the usage of the terms "Cybernetics" and "Control" in scientific publications (paper titles) indexed by Scopus. Clearly, these terms appear interdisciplinary and widespread in many branches of modern science.

---

[13]Cybernetical physics is a science studying physical systems by cybernetical methods. Owing to the maturity of physical objects modeling (in the sense of duration and depth), today the results in this field can be stated as general and well grounded laws, see [59, pp. 38–40] and below.

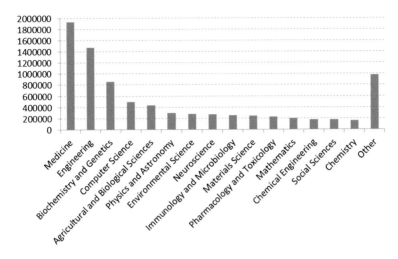

Fig. 1.3 The usage of the term "Control" by scientific branches in paper titles indexed by Scopus

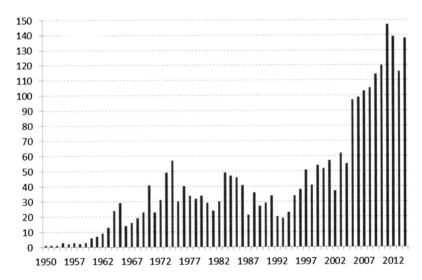

Fig. 1.4 The usage of the term "Cybernetics" by years in publications indexed by Scopus

Figures 1.4 and 1.5 illustrate the usage of the terms "Cybernetics" and "Control" by years in scientific publications indexed by Scopus.

And finally, Fig. 1.6 shows the usage of the terms "Cybernetics" and "Control" by years in the texts of publications indexed by Google Scholar. The dip observed in recent years can be explained by indexing delays.

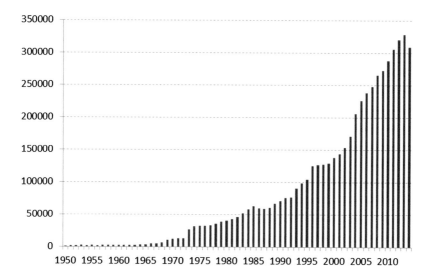

Fig. 1.5 The usage of the term "Control" by years in publications indexed by Scopus

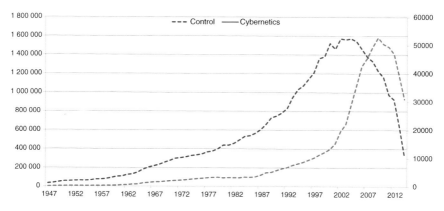

Fig. 1.6 The usage of the terms "Cybernetics" and "Control" by years in the texts of publications indexed by Google Scholar

1.2   Cybernetics of Cybernetics and Other Types of Cybernetics

In addition to Wiener's classical cybernetics, the last 50+ years yielded other types of cybernetics declaring their connection with the former and endeavoring to develop it further.

No doubt, the most striking phenomenon was the appearance of *second-order cybernetics* (cybernetics of cybernetics, metacybernetics, new cybernetics; here

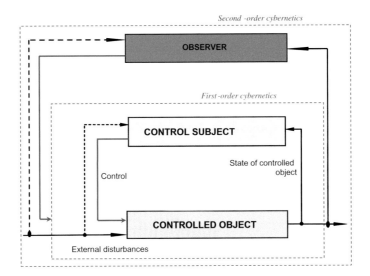

Fig. 1.7 First- and second-order cybernetics

"order" corresponds to "reflexion rank"). Cybernetics of cybernetic systems is
associated with the names of M. Mead, G. Bateson and H. Foerster and puts its
emphasis on the role of subject/observer performing control[14] [20, 54, 55, 81, 128]
(see Fig. 1.7). The central concept of second-order cybernetics is an *observer* as a
subject refining the subject from the object (indeed, any system is a "model"
generated from reality for a certain cognitive purpose and from some point of
view/abstraction).

H. Foerster noted that "a brain is required to write a theory of a brain. From this
follows that a theory of the brain, that has any aspirations for completeness, has to
account for the writing of this theory. And even more fascinating, the writer of this
theory has to account for her or himself. Translated into the domain of cybernetics;
the cybernetician, by entering his own domain, has to account for his or her own
activity. Cybernetics then becomes cybernetics of cybernetics, or second-order
cybernetics." [55].

In contrast to Wiener's cybernetics, second-order cybernetics possesses *the
conceptual-philosophical character* (for a mathematician or engineer, it is demon-
strative that all publications on second-order cybernetics contain no formal models,
algorithms, etc.). In fact, this type of cybernetics "transmits" the complementarity
principle (with insufficient grounds) from physics to all other sciences, phenomena
and processes. Moreover, a series of works postulated that any system must
have positive feedback loops amplifying positive control actions (e.g., see [124]).

---

[14]Such approach has been and still is conventional for theory of control in organizations (e.g., see
Fig. 1.4.15 in [131, 159]).

But any expert in control theory knows the potential danger of such loops for system stability!

The "biological" stage in second-order cybernetics is associated with the names of Maturana and Varela [125, 126, 210] and their notion of *autopoiesis* (self-generation and self-development of systems). F. Varela underlined that "first-order cybernetics is the cybernetics of observed systems; second-order cybernetics is the cybernetics of observing systems." The latter focuses on feedback of a controlled system and an observer.

Therefore, the key terms of second-order cybernetics are recursiveness, self-regulation, reflexion, autopoeisis. For a good survey of this direction, we refer to [116].

Asaro [13] believed that there exist three interpretations of cybernetics (actually, we have mentioned the first two above):

1. the narrow interpretation, i.e., a science studying feedback control;
2. the wide interpretation, i.e., "cybernetics is all the things, and we live in the Century of Cybernetics";
3. the intermediate (epistemological) interpretation, i.e., second-order cybernetics (an emphasis on feedback of a controlled system and an observer).

However, the historical picture has appeared much more colorful and diverse, not confining to the second order–see Fig. 1.8.

Some authors adopt the terms "*third-order cybernetics*" (social autopoeisis; second-order cybernetics considering autoreflexion) and "*fourth-order cybernetics*" (third-order cybernetics considering observer's system of values), but they are conceptual and still have no generally accepted meanings (e.g., see a discussion in [31, 95, 121, 122, 140, 206, 207]).

For instance, V. Lepsky wrote: "Third-order cybernetics can be formed basing on the thesis "from observing systems to self-developing systems." In this case, control is gradually transformed into a wide spectrum of support processes for system self-development, namely, social control, stimulation, maintenance, modeling, organization, "assembly/disassembly" of subjects and others." [118, p. 7793].

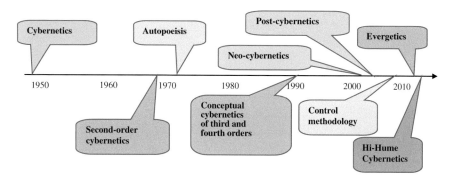

Fig. 1.8   The ontogenesis of cybernetics: different types of cybernetics

We point out other directions (see Table 1.1 and Chap. 2 of the book):

- *homeostatics* (Yu. Gorsky and his scientific school), a science studying con-
  tradictions control for the sake of maintaining the permanency of processes,
  functions, development trajectories, etc. [71];
- *neo-cybernetics* (B. Sokolov and R. Yusupov), an interdisciplinary science
  which elaborates a methodology of stating and solving analysis and synthesis
  problems of intelligent control processes and systems for complex
  arbitrary-nature objects [191, 192];
- *neo-cybernetics* (S. Krylov) [111];
- *new cybernetics*, *post-cybernetics* (G. Tesler), a fundamental science about
  general laws and models of informational interaction and influence in processes
  and phenomena running in animate, inanimate and artificial nature [199].
  Interestingly, K. Kolin had proposed almost a same definition to informatics
  20 years before Tesler, see [101];
- *evergetics* (V. Vittikh), a value-oriented science about control processes in a
  society, which focuses on problem situations for a group of heterogeneous
  actors with different viewpoints, interests and value preferences [212]. In other
  words, evergetics can be defined as third-order cybernetics for interacting
  control subjects. According to Vittikh's fair remark, in everyday social life
  control processes will be realized by the "tandem" of common and professional

Table 1.1 Different types of cybernetics

Type	Authors	Period
Cybernetics	N. Wiener W. Ashby S. Beer	The 1948–1950s
Second-order cybernetics	M. Mead G. Bateson H. Foerster	The 1960–1970s
Autopoiesis	H. Maturana F. Varela	The 1970s
Homeostatics	Yu. Gorsky	The 1980s
Conceptual cybernetics of third and fourth orders	V. Kenny R. Mancilla S. Umpleby	The 1990–2010s
Neo-cybernetics	B. Sokolov R. Yusupov	The 2000s
Neo-cybernetics	S. Krylov	The 2000s
Third-order cybernetics	V. Lepsky	The 2000s
New cybernetics, post-cybernetics	G. Tesler	The 2000s
Control methodology	D. Novikov	The 2000s
Evergetics	V. Vittikh	The 2010s
Subject-oriented control in noosphere (Hi-Hume cybernetics)	V. Kharitonov A. Alekseev	The 2010s

control experts (theoreticians): the former face concrete problem situations in daily routine and acquire conventional knowledge (in the sense of H. Poincare) on the situation and define directions of its control, whereas the latter create necessary methods and means for their activity. Involvement of "common" people into social control processes is an important development trend of control science.

– *subject-oriented control in noosphere,* the so-called *Hi-Hume Cybernetics* (V. Kharitonov and A. Alekseev), a science mostly considering subjectness and subjectivity of control [79].

It is possible to introduce the notion of *"fifth-order cybernetics"* as fourth-order cybernetics considering the mutual reflexion of control subjects [158] making coordinated decisions, etc. Note that all types of cybernetics in Table 1.1 are conceptual, i.e., absorbed by Cybernetics.

The observed variety of the approaches claiming (explicitly or implicitly) to be a new mainstream in classical cybernetics development seems natural, as reflecting the evolution of cybernetics. With the lapse of time, certain approaches will be further developed, others will stop growing. Of course, it is extremely desirable to obtain a general picture with integration, generalization and joint positioning of all existing approaches or most of them (see the Conclusion).

1.3 Achievements and Disillusions of Cybernetics

Cybernetics has been always assigned a wide range of assessments by experts and everymen (at least, since the middle of the 1960s), from "cybernetics has discredited itself against all expectations and ceased to exist by today" to "cybernetics covers all the things."[15] As ever, the truth is a golden mean.

Some doubts in the existence of cybernetics "today" and arguments witnessing for it (e.g., see [170, 191, 192]) have been repeatedly stated starting from the middle of the 1980s. Here are some quotations:

– "As a scientific discipline, cybernetics still exists but its claims for the role of some all-embracing control science disappeared." [135];
– "We have to acknowledge that general cybernetics has failed to form a scientific discipline." "… It is difficult to find a specialist identifying himself as a cybernetician." [173];
– "Today the term "cybernetics" is mentioned here, there and everywhere on and off the point." [136].

Such opinions are only partially correct. Cybernetics was born in the middle of the 1940s as the science of "control and communication in the animal and the machine," or even as the science of GENERAL control laws (recall the definitions

---

[15]This also applies to systems theory and systems analysis, see Chap. 4.

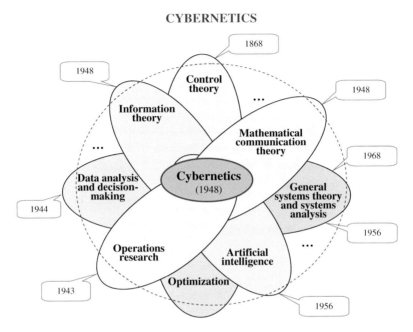

Fig. 1.9 The composition and structure of cybernetics

of cybernetics and Cybernetics above and Fig. 1.9). The triumph of cybernetics in the 1950–1960s, namely, the appearance of technical, economic, biological and other types of cybernetics, their close connections with operations research, mathematical control theory, as well as intensive application of its results in design and refinement of technical and information systems, created the illusion of universalism and the illusion of inevitable rapid progress of cybernetics in future. Nevertheless, in the early 1970s the development of cybernetics slowed down, its integral flow was decomposed into numerous partial subflows and "lost in details": the number of scientific directions[16] (see Fig. 1.9) increased and each of them continued further development, but general regularities were almost not identified and not systematized. Actually, cybernetics had rapid growth owing to its components, but Cybernetics stood still.

Concerning Fig. 1.9 and similar ones (see Figs. 3.2, 3.3 and A.2), the author addresses esteemed readers with an appeal to acknowledge that any ideas about the correlation of sciences and their branches *are* "*egocentric*"–a scientist places its own ("favorite") science "in the core." Moreover, any scientific branch or scientific school *hyperbolizes* its achievements and capabilities. Such subjectivism seems natural, and a real picture can be always reconstructed with appropriate corrections to it.

---

[16]Exactly scientific directions, i.e., sciences, group of sciences and application domains.

Another argument: since the 1950s, the mankind has been observing the "exponential" growth of technological innovations and the same growth of scientific publications, parallel to *sciences differentiation* (N. Wiener wrote: "Since Leibniz there has perhaps been no man who has had a full command of all the intellectual activity of his day." [221, p. 43]). An interesting paradox: over this period, the number of researchers, scientific papers, journals and conferences has been increasing, but almost without the appearance of revolutionary fundamental scientific discoveries "clear to everymen." Fundamental science "has passed ahead of" technologies and its groundwork is now implemented in new technologies. Yet, intensive development of fundamental science cannot be stimulated without an explicit mass "demand" from technologies.

In the era of cumulative differentiation of sciences, cybernetics has been remaining a striking example of the synergetic effect, i.e., a successful attempt to integrate different sciences, to search their common language and regularities. Unfortunately, it is one of the last examples: modern fashionable "convergent sciences" (NBICS: nano-, bio-, informational, cognitive and humanitarian social sciences) have still not realized themselves in this sense. Indeed, the widespread "*interdisciplinarity*" is rather an advertising umbrella brand or a real "junction" of two or more sciences. Genuine **Interdisciplinarity** must operate common results and regularities of several sciences.

As an epistemological digression, note that the *dialectic spiral* "from partial to generalizations, from generalizations to new partial results" is characteristic for any-scale theory, i.e., from partial (yet, integral) direction of investigations[17] to full-scale scientific research (see Fig. 1.10 imported from [148]). Wiener's ideas about the general regularities of control and communication in different-nature systems were the result of generalizing some (of course, not all!) achievements of automatic control theory, communication theory, physiology and a series of other sciences of that time. Wiener's cybernetics with the key concepts of feedback (causality), homeostasis and others spurred new results in control, informatics and other sciences.

Thus, the "romantic" period (see the Introduction) was followed by the period of rapidly obtained results, *ergo* by the growing expectations. Those expectations were not necessarily professional. Cybernetics became fashionable and many authors started its popularization.[18] Sometimes, the number of popularizers even exceeded

---

[17]For instance, an efficient solution method for a certain class of control problems becomes applicable to problems in adjacent fields (e.g., communication, production, etc.). Thereby, this method is "transferred" from control theory to cybernetics. And then, it can be an asset of applied mathematics, i.e., a "spanner" for experts in various fields (whenever studied systems satisfy its initial requirements).

[18]Actually, the first popularizer was N. Wiener himself. Later, he mentioned that the appearance of the book [221] in a moment transformed him from a working scientist with a definite authority in his research field into some public figure. That was pleasing, but also had negative consequences, as henceforth N. Wiener was obliged to maintain business relations with various scientific groups and take part in a movement which rapidly gained in scope so that he could not even cope with it.

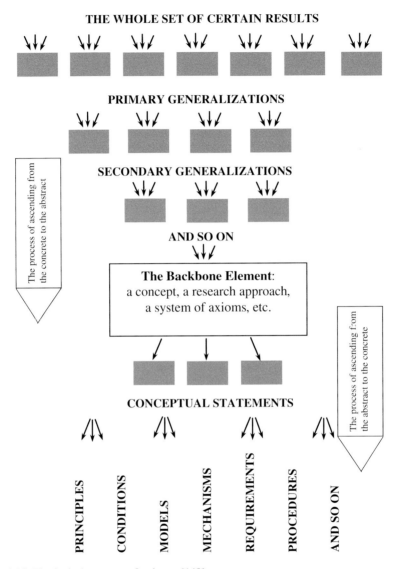

Fig. 1.10 The logical structure of a theory [148]

the number of professionals (for the sake of justice, we emphasize that professionals realized not all their expectations). A. Kolmogorov was right saying that "I do not belong to great enthusiasts of all rich literature on cybernetics published today and see numerous exaggerations (on the one part) and much oversimplification (on the other part) in it." [104].

Perhaps, such situation is typical for the development of scientific branches and directions. There exist many examples of failed expectations originally created and

maintained by dilettantes. For instance, the terminology of rather fruitful independent sciences such as nonlinear dynamics and synergetics [45, 77, 120, 175, 194]
(attractors, bifurcations, etc.) is often employed by humanists for constructing a
scientific entourage for the outsiders. Fuzzy set theory, artificial neural networks,
genetic algorithms and many other scientific fields have already passed or are now
facing a crisis due the collapse of corresponding overrated expectations.

Consider the following groups of subjects:

- researchers focused on cybernetics proper;
- researchers representing adjacent (component) sciences;
- popularizers of cybernetics (mass media or dilettantish "researchers" interpreting the results of others[19]);
- authorities ("politicians") and potential users of applied results ("customers") in
  business structures.

The failed expectations for cybernetics caused disillusions of all these groups.
Answering to the question "Where are the results?", experts in cybernetics parried:
"We work as hard as possible; all promises were given by popularizers and they
must bear the responsibility." Due to their sound "jealousy" to cybernetics, the
representatives of adjacent sciences replied "The things are going well with us"[20]
(really, many "components" of cybernetics such as control theory, informatics and
others were quite successful, see Fig. 1.9). Popularizers infrequently feel pangs of
conscience[21] and can always note: "We are not experts, we were deluded." With the
course of time, politicians also felt definite pessimism over cybernetics, i.e., particularly due to the attitude towards cybernetics in the early 1950s in the USSR, the
Chilean experiments of S. Beer's team (implementation of cybernetical ideas and
approaches in real economy management) and V. Glushkov's unrealized intentions
to deploy all-embracing computer-based centralized planning in the USSR.

No guilty persons found, something failed, and that's it. Actually, the situation is
not so bad as it seems to be. First, **cybernetics is rather efficient as an integrative
science**: its components have been and will be developing for years, while a unique
look and a holistic picture covering a whole group of sciences is surely needed (see
Chap. 2). Reflexion with respect to disillusions and their reasons is anyway useful.

Second, for several decades cybernetics was considered as a "magic lamp"
throwing the light on the correct structure of any subject domain and systematizing
its organization (N. Moiseev noted that cybernetics defines "a thinking standard"
[136]). In many cases (technical systems, numerous results in biology and

---

[19]Such "researchers" exist in any science, especially in and around intensively developing ones.

[20]In fact, valuable results in automatic control theory, statistical communication theory, etc. were
followed by some recession (quite naturally, see Fig. 5.11).

[21]During his speech at 1962 IFIP Meeting, USSR representative A. Dorodnitsyn suggested two
terms for the glossary of information processing, namely, "Cybernetics active" and "Cybernetics
talkative."

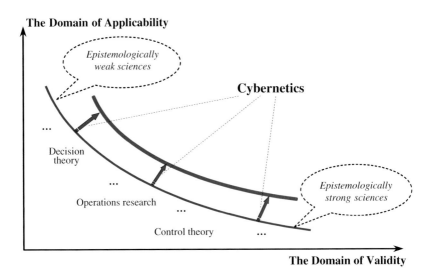

Fig. 1.11 The "principle of uncertainty"

economics, etc.), the hopes were justified, inducing higher expectations. Any synthetic science including cybernetics represents not a "lamp," but a "lens" properly focusing rays (scientific and applied results) from a "source" (concrete sciences): a lens gives no light, but acts as a light converter.

The main *problem of cybernetics* as a "lens" consists in the following. Except the founding fathers of classical cybernetics (N. Wiener, W. Ashby and S. Beer), **just a few researchers studied Cybernetics deeply and professionally** endeavoring to reveal, formulate and develop its general laws (see Chap. 3), despite the huge growth of knowledge in adjacent sciences within the past decades. (A new turn of appreciable generalizations took no place, see Fig. 1.10). Moreover, the interdisciplinarity of cybernetics (multiple subjects and methods of study[22]) testified to its "fuzziness." Contrariwise, Cybernetics is a more holistic science with its own subject–general regularities of control and communication. Therefore, experts and specialists **should pay their attention and apply every effort to develop Cybernetics**!

Concluding this section, recall the "*principle of uncertainty*" described in [149]: epistemologically weak sciences introduce the minimal constraints (or no constraints at all) and obtain the fuzziest results. Contrariwise, epistemologically strong sciences impose many limiting conditions, involve scientific languages, but yield more precise (and well-grounded) results. However, the field of their application appears rather narrowed (i.e., clearly bounded by these conditions). In other words, the current level of science development is characterized by certain mutual constraints imposed on results "validity" and results applicability, see Fig. 1.11. That is,

---

[22]In this sense, interdisciplinarity is rather a negative feature.

the "product" of the domains of results applicability and validity does not exceed a constant (increasing the value of a "multiplicand" reduces the value of another "multiplicand").

But this regularity applies only to a current development level of a corresponding science. The presence of generalizations (the main role of Cybernetics!) extends the boarders by shifting the curve to the right and top (see Fig. 1.11). As a result, some progress is achieved in the both domains.

Chapter 2
Cybernetics, Control Philosophy and Control Methodology

Having reached a certain level of epistemological maturity, scientists perform "reflexion" by formulating general laws in corresponding scientific fields, i.e., create *metasciences* [149, 152]. On the other part, any "mature" science becomes the subject of philosophical research. For instance, the philosophy of physics appeared at the junction of the 19th century and the 20th century as the result of such processes.

Originated in the 1850s, research in the field of *control theory* led to the appearance of other metasciences, i.e., *cybernetics* and *systems analysis*. Moreover, cybernetics quickly became the subject of philosophical investigations (e.g., see [20, 50, 54, 87, 97, 98, 126, 176, 207, 208]) conducted by "fathers" of cybernetics and professional philosophers.

The 20th century was accompanied with the rapid progress of *management science* [38, 131, 157] as a branch of general control theory studying practical control in *organizational systems*. By the beginning of the 2000s, management science has engendered *management philosophy*. Books and papers entitled "Management Philosophy" and "Control Philosophy" appeared exactly at that times (for instance, see references in [152]); as a rule, their authors represented professional philosophers. Generally speaking, one may acknowledge the long-felt need for more precise mutual positioning of philosophy and control, methodology and control, as well as analysis of general laws and regularities of complex systems functioning and control.

2.1 Control Philosophy

Historically (and similarly to the subjects of most modern sciences), control problems analysis was first the prerogative of *philosophy*. R. Descartes was used to say, "Philosophy is like a tree whose roots are metaphysics and then the trunk is physics. The branches coming out of the trunk are all the other sciences."

© Springer International Publishing Switzerland 2016
D.A. Novikov, *Cybernetics*, Studies in Systems, Decision and Control 47,
DOI 10.1007/978-3-319-27397-6_2

R. Mirzoyan felt rightly that, on the basis of historical and philosophical analysis, first control/management theorists were exactly philosophers [135]. Confucius, Lao-tzu, Socrates, Platon, Aristotle, N. Machiavelli, T. Hobbes, I. Kant, G. Hegel, K. Marx, M. Weber, A. Bogdanov—this is a short list of philosophers that laid down the foundations of modern control theory for the development and perfection of managerial practice.

Consider Fig. 2.1 [152] illustrating different connections between the categories of *philosophy* and *control*; they are treated in the maximal possible interpretation (philosophy includes ontology, epistemology, logic, axiology, ethics, aesthetics, etc.; control is viewed as a science and a type of practical activity). We believe that the three shaded domains in Fig. 2.1 are the major ones.

Presently, concrete control problems are no more the subject of philosophical analysis. Philosophy (as a form of social consciousness, the theory of general principles of entity and cognition, human attitude to the reality, as the science of universe laws of natural development) studies GENERAL problems and regularities separated out by experts in certain sciences.

V. Diev believed that control philosophy is "a system of generalizing philosophical judgments about the subject and methods of control, the place of control

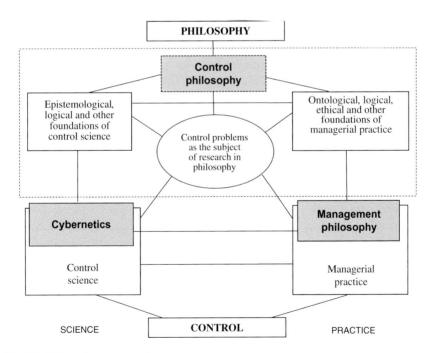

Fig. 2.1 Philosophy and control

among other sciences and in the system of scientific cognition, as well as about its cognitive and social role in a modern society." [50, p. 36].

One can define *control philosophy* as a branch of philosophy connected with comprehension and interpretation of control processes and control cognition, studying the essence and role of control [152]. Such meaning of the term "control philosophy" (see the dashed-line contour in Fig. 2.1) has a rich internal structure and covers epistemological research of control science, the analysis of logical, ontological, ethical and other foundations (both for control science and managerial practice).

Cybernetics (with capital C, as a branch of control science, studying its most general theoretical regularities). According to V. Diev, "… for many scientific disciplines, there exists a range of problems related to their foundations and traditionally referred to as the philosophy of a corresponding science. Control science follows this tradition, as well." [50, p. 36]. Foundations of control science also include general regularities and principles of efficient control representing the subject of Cybernetics (see Chap. 3).

In the 1970–1990s, against the background of first disillusions of cybernetics, the only bearers of canonical cybernetic traditions were philosophers (!), whereas experts in control theory lost their confidence in ample opportunities of cybernetics.

Things can't carry on as they are. On the one hand, philosophers vitally need knowledge of the subject (actually, the generalized knowledge). In this context, V. Il'in mentioned that "philosophy represents second-rank reflexion; it provides theoretical grounds to other ways of spiritual production. The empirical base of philosophy consists in specific reflections of different types of cognition; philosophy covers not the reality itself, but the treatment of reality in figurative and category-logical forms." [87].

On the other hand, experts in control theory need "to see the wood for the trees." Hence, one can hypothesize that **Cybernetics must and would play the role of control "philosophy"** (here quotation marks are crucial!) as a branch of control theory, studying its most general regularities. Here the emphasis should be made on constructive development of control philosophy, i.e., on formation of its content through obtaining concrete results (probably, first partial results and then general ones). Reflexion can be continued by considering *cybernetics philosophy*, and so on.

The book [152] briefly analyzed the correlation of control philosophy (as a branch of philosophy studying general problems of control theory and practice), Cybernetics (as a branch of control science generalizing the methods and results of solving theoretical problems of control) and management "philosophy" (as a branch of control science generalizing the experience of successful managerial practice), see Fig. 2.2.

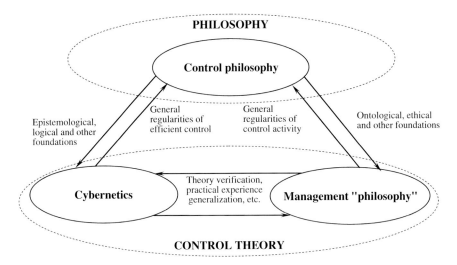

Fig. 2.2 Control philosophy, Cybernetics and management "philosophy"

2.2   Control Methodology

Methodology is the theory of activity organization [148, 149]. Accordingly, the subject of methodology is organization of an activity (an *activity* is a purposeful human action).

Control activity represents a certain type of practical activity. *Control methodology* is the theory of organization of control activity, i.e., the activity of a control subject [152]. Whenever a control system incorporates a human being, control activity becomes *activity on activity organization*. Control theory puts its emphasis on the interaction of control subject and controlled object (the latter can be another subject), see Fig. 2.3. At the same time, control methodology explores the activity of a control subject, *ergo* has-to-be-included in Cybernetics.

The development of control methodology formulated *the structure of control activity* (see Fig. 2.4) and identified the structural components of control theory [152].

A *theory* is an organizational form of scientific knowledge about a certain set of objects, representing a system of interconnected assertions and proofs and containing methods of explanation and prediction of phenomena and processes in a given *problem domain*, i.e., of all phenomena and processes described by this theory. First, any scientific theory consists of interrelated structural elements. Second, any theory includes in its initial basis a *backbone element* [148].

The backbone element of control theory (for social systems, organizational systems and other interdisciplinary systems) is the category of organization[1];

---

[1]Note that "organism" and "organization" are paronyms.

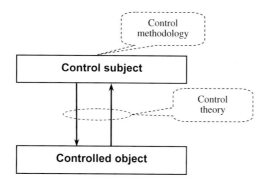

Fig. 2.3 Control methodology and control theory

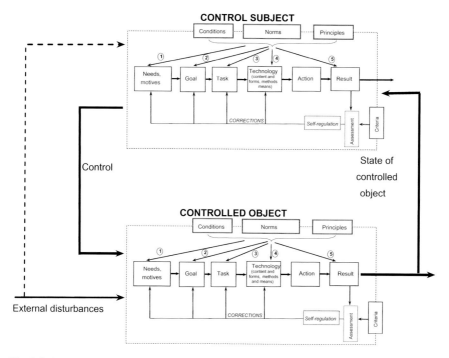

Fig. 2.4 Structural components of control activity

indeed, control is the process of organizing which leads to the property of good
organization as a property in a controlled system (see the Conclusion).

The structural *components of control theory* (see Fig. 2.5) are:

– control tasks;
– scheme of control activity;
– conditions of control;

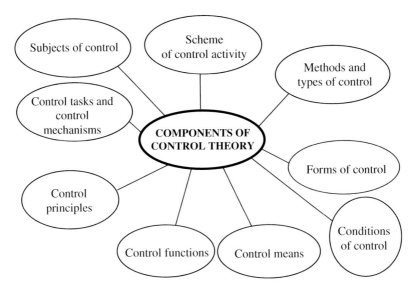

Fig. 2.5 Components of control theory

- types of control;
- subjects of control;
- methods of control;
- forms of control;
- control means;
- control functions;
- factors having an impact on control efficiency;
- control principles;
- control mechanisms.

They are considered in detail in [152].

The foundations of control methodology, the characteristics of control activity, its logical and temporal structures, as well as the *structure of control theory* (as a set of stable relations among its components) are discussed in [152, 157, 158].

Chapter 3
Laws, Regularities and Principles of Control

Among important subjects studied by Cybernetics, we mention laws, regularities and principles of complex systems functioning and control.

Laws, regularities and principles. According to Merriam-Webster Dictionary, a *principle* is:

1. a basic statement of a certain theory, science, etc.; a guidance idea or a basic rule of an activity;
2. an internal belief or view of something, which defines norms of behavior;
3. a key feature in the structure of a mechanism, a device or an installation.

Let us adhere to the first interpretation of the term "principle"; thus, *control principles* will be understood as the rules of control activity. We will also address its third interpretation as a key feature in system structure.

A *law* is a permanent cause-and-effect relation of phenomena or processes.

A *law* is a necessary, essential, stable and repetitive relation among phenomena.

In contrast to laws, regularities are not compulsory; principles can be treated as strict imperatives or desirable properties.

There exists *a hierarchy of laws and principles* (see Fig. 3.1): philosophical laws are most general; the next level is occupied by more "partial" logical and other general scientific laws and principles (including the ones of cognition and practical activity, see [148, 149]); and finally, laws, regularities and principles of specific sciences appear least general (on the one hand, control theory as a science possesses its own laws and principles; on the other hand, it employs laws and principles of other sciences relating to a controlled object).

Which are general and accepted control laws? Unfortunately, today one would hardly provide an exhaustive answer.

First, we should distinguish between two well-established interpretations of the term "control law." The general interpretation has been already given. The narrow interpretation states that a control law is a relationship or a class of relationships between control actions and available information on the state of a controlled system (i.e., the law of proportional control, proportional-integral control, etc.). We are concerned with the general interpretation.

© Springer International Publishing Switzerland 2016
D.A. Novikov, *Cybernetics*, Studies in Systems, Decision and Control 47,
DOI 10.1007/978-3-319-27397-6_3

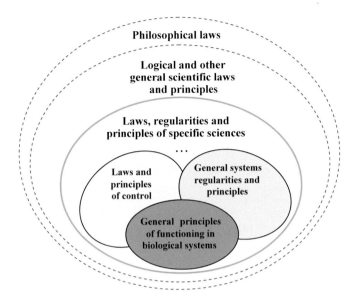

Fig. 3.1 The hierarchy of laws, regularities and principles

Second, it would seem that many laws of modern control science are not control laws in the above general sense. For instance, feedback control is widespread in control theory but does not appear universe. Indeed, there exists programmed control, and other types of control involving no direct information on the current state of a controlled system.

Third, "control laws" mentioned in scientific literature (such as the presence of a goal, the presence of a feedback, etc.) are rather control principles or control regularities than control laws[1] (see below). We consider well-known control laws.

General Control Laws (Regularities)

1. *The law of goal-directedness*—any control has a *goal*;
2. *The law of requisite variety* (sometimes called *the adequacy principle* stated by W. Ashby [14])—the variety[2] of a controller must be adequate to the variety of a controlled object.[3] In [19] variety was treated as complexity, and the law of

---

[1]The following opinion is also cultivated in scientific community. Being a language, mathematics has no inherent laws (e.g., in contrast to natural sciences); similarly, control theory as a general descriptive language of control processes operates no inherent laws till a class of controlled objects is specified.

[2]A quantitative characteristic of a system determined as the number of admissible states or the logarithm of this quantity.

[3]The law of requisite variety should be given a more precise definition: the variety of a controller must be adequate to the variety of a CONTROLLED SUBJECT reflecting the goal aspects of a

requisite variety was formulated as *the law of requisite complexity*. Ashby himself believed that "every law of nature is a constraint" [14].

3. *The law of emergence* (synergy) is the main law of systems theory. It claims that "the whole is greater than the sum of its parts" (Aristotle); in other words, the properties of a system are richer than the "sum" of the properties of its elements. W. Ashby believed that, the greater is a system and the bigger are the existing difference between the sizes of the whole and its parts, the higher is the probability that the properties of the whole differ appreciably from the properties of its parts.

4. *The law* of *external complementarity* was suggested by S. Beer (the so-called third principle of cybernetics): any control system needs a black box, i.e., certain reserves for compensating the disregarded impact of external and internal environments (actually, this idea underlies robust control).

5. *The law or principle of feedback* (cause-and-effect relations)—see below.

6. *The law of optimality*—a control action must be "best" in the sense of goal attainment under existing constraints.[4] L. Euler wrote: "Since the fabric of the universe is most perfect and the work of a most wise Creator, nothing at all takes place in the universe in which some rule of maximum or minimum does not appear." On the other part, Yu. Germeier thought that by observing a certain behavior of a system one can a posteriori construct a functional optimized by this behavior [64]. The law of optimality does not imply that all real systems are optimal, i.e., have the maximum efficiency; rather, it serves as a norm for artificial/control systems designers.

The above-mentioned principles are often accompanied by the principles of *causality*, *decomposition* (*analysis*), *aggregation* (*synthesis*), *hierarchy*, *homeostasis*, *consistency* (prior to control design, consider the problems of observability, idenfiability, controllability including stability), *adaptability*, and others.

Some authors (e.g., see [80, 174] and a survey in [152]) proposed their own laws, regularities and principles of cybernetics, control and development. First, many principles stated in literature are disputable, as representing the examples of

---

(Footnote 3 continued)

controlled object. Indeed, one would hardly imagine a "controller" with greater variety than a human being.

[4]Optimization consists in seeking for best alternatives among a set of admissible ones under given constraints (optimal alternatives). In this phrase each word is important. "Best" means the presence of a criterion (or several criteria) and a way (several ways) to compare alternatives. It is crucial to take into account existing conditions and constraints: their variation possibly leads to a situation when other alternatives appear best under a same criterion (same criteria). The notion of optimality has received a rigorous and exact representation in different mathematical sciences, has firmly entrenched in practical design and exploitation of technical systems, has played a prominent role in formation of modern systems ideas. Moreover, this notion is widespread in administrative and public practice and is known to almost everyone. Obviously, aspiration for increasing the efficiency of any purposeful activity has found its expression, a clear and intelligible form in the idea of optimization.

an unadapted groundless transfer and/or "generalization" of results. For instance, V. Pareto established empirically that 20 % of population own 80 % of capital in the world [164]. Nowadays, *the Pareto principle* (also known as "the 80/20 principle" or "the beer law"[5]) is formulated as a universal natural law without proper substantiation:

– 20 % of efforts yield 80 % of a result;
– 80 % of company's stocking cost corresponds to 20 % of its product types;
– 80 % of company's sales income is made by 20 % of its customers;
– 80 % of problems are created by 20 % of causes;
– 20 % of working time is spent on 80 % of work;
– 80 % of work is performed by 20 % of employees, and so on.

Another example concerns *the principle of harmony*. Using the proportions established by L. da Vinci (the *golden section*) and the well-known properties of the Fibonacci sequence, one postulates the corresponding ratio of other indicators (e.g., the number of employees, wages, budget articles, and so on).

Such "principles" and their apologists can be treated with a smile, as both the former and the latter have no attitude to science proper.

Second, all researchers (!) have not stated any *enumeration bases* for principles and laws suggested by them. This fact testifies to their possible non-universalism, as well as to incomplete enumeration, its weak soundness, possible internal inconsistency, etc.

And third, the list of laws, regularities and principles should be extended and systematized.

As an illustration, consider some principles and laws of control and functioning of complex systems proposed by different authors.

Principles of Complex Systems Functioning [86, pp. 60–67]

1. *The principle of reactions*—responding to an external influence, a system reinforces processes to compensate it (the Le Chatelier–Brown principle imported from physics and chemistry).
2. *The principle of system cohesion*—a system's form is maintained by a balance, static or dynamic, between cohesive and dispersive influences. The form of an interacting set of systems is similarly maintained.
3. *The principle of adaptation*—for continued system cohesion, the mean rate of system adaptation must equal or exceed the mean rate of changes of environment (the response times obey the reverse rule).
4. *The principle of connected variety*—interacting systems stability increases with variety, and with the degree of connectivity of that variety within the environment.

---

[5]20 % of people drink 80 % of beer.

5. *The principle of limited variety*—variety in interacting systems is limited by the available space and the minimum degree of differentiation.
6. *The principle of preferred pattern*—the probability that interacting systems will adopt locally-stable configurations increases both with the variety of systems and with their connectivity.
7. *The principle of cyclic progression*—interconnected systems driven by an external energy source will tend to a cyclic progression in which system variety is generated, dominance emerges to suppress the variety, the dominant mode decays or collapses, and survivors emerge to regenerative variety.

According to [156], most well-known principles and laws of functioning of complex (in the first place, biological) systems are exactly regularities or hypotheses. To explain this statement, consider **PRINCIPLES OF BIOLOGICAL SYSTEMS**[6] **FUNCTIONING** which are also the subject of Cybernetics (see surveys in [10, 156]).

1. **The principle of least action**. A dynamic system moves from an initial configuration to a final configuration in a specified time along a trajectory which minimizes the action (a functional of the trajectory). Actually, this principle coincides with the law of optimality pioneered in physics in the 1790–1800s.
2. **The principle of the permanent inequilibrium** (E. Bauer, 1935). The living and only the living systems are never in an equilibrium, and, on the debit of their free energy, they continuously invest work against the realization of the equilibrium which should occur within the given outer conditions on the basis of the physical and chemical laws [21, p. 43] (see the principle of reactions).
3. **The principle of simplest structure** (N. Rashevsky, 1943). A concrete structure of a living system which exists in nature is the simplest among all structures being able to perform a given function or a set of functions [178].
4. **The principle of feedback** (see also *the principle of functional systems* by P. Anokhin [9]). In this context, we have to mention his principle of anticipatory reflection or reality: "One universal regularity was formed during the adaptation of the organisms to the environment, which was further developed during the whole period of evolution of living organisms: the highest order of speed for the reflection of the low speed deployment of the events of the real world." [7]. A complex adaptive system responds not to an external influence as a whole, but "to the first chain of a repeated series of external influences." [7]. Practical realizations of the principle of feedback have a long history—from several mechanisms in Egypt (Ctesibius' water clock, the 2nd–3rd century B.C.) to perhaps first feedback usage in Drebbel's thermostat (1572–1633), Polzunov's water-level float regulator (1765) and Watt's steam engine governor (1781), Jacquard's loom with program control (1804–1808), etc.

---

[6]Interestingly, the overwhelming majority of these principles were formulated in the 1940–1960s.

The pioneering fundamental works on mathematical control theory were published by J. Maxwell [127] and I. Vyshnegradsky [216].[7] Actually, the first general systematic analysis of feedback was performed by P. Anokhin [8], later jointly by A. Rosenblueth et al. [181] and, in the final statement, by N. Wiener [221]. For justice' sake, note that feedback was studied and used in electrical engineering in the 1920s.

5. **The principle of least interaction** (I. Gelfand and M. Tseitlin, 1962 [60]). Nerve centers aspire to achieve a situation when *afference* (informational and control flows and signal transmitted in central nervous system) is minimal. In other words, a system functions rationally in some external environment if it seeks to minimize interaction with the environment [202].

6. **The principle of brain's stochastic organization** (A. Kogan, 1964 [100]). Each neuron has no independent function, i.e., is a priori not responsible for solution of a concrete task; all tasks are distributed randomly.

7. **The principle of hierarchical organization** (particularly, information processing by brains), see the works of N. Amosov [5], W. Ashby [15], N. Bernshtein [24], and G. Walter [218]. Achieving a whole goal is equivalent to achieving the set of its subgoals.

8. **The principle of adequacy** (W. Ashby, 1956 [14], Yu. Antomonov [10] and others). For effective control the complexity of the controller (dynamics of its states) must be adequate to the complexity (rate of change) of controlled processes. In other words, the "capacity" of the controller defines the absolute limit of control regardless of the capabilities of the controlled system (see the law of requisite variety above).

9. **The principle of probabilistic prediction** in actions design (N. Bernshtein, 1966) [24]. The world is reflected in two models, viz., *the model of the desired future* (probabilistic prediction based on previously accumulated experience) and the model of the backward (which explicitly reflects the observed reality).

10. **The principle of necessary degree of freedom selection** (N. Bernshtein, 1966). Initially, learning involves more degrees of freedom of a learned system than is actually required for learning [24]. During learning, the number of "involved" variables decreases as inessential variables are "eliminated" (compare this principle with the phenomena of generalization and concentration of nervous processes—I. Pavlov, A. Ukhtomskii, P. Simonov, and others).

11. **The principle of determinism destruction** (H. Foerster, Yu. Antomonov [10, 55] and others, 1966). To achieve a qualitatively new state and to increase the level of system organization, it is necessary to destroy (rearrange) the existing deterministic structure of connections among system elements, which was formed by the previous experience.

12. **The principle of requisite variety** (W. Ashby, 1956). This principle (see above) is close to the principle of adequacy [14].

---

[7]The first course of lectures entitled "Theory of direct-action regulators" by D. Chizhov appeared in Russia in 1838.

13. **The principle of natural selection** (S. Dancoff, 1953). In systems becoming efficient due to natural selection, the variety of mechanisms and capacity of information transmission channels does not appreciably exceed the minimum level required [48].
14. **The principle of deterministic representation** (J. Kozielecki, 1979 and others). Modeling of decision-making by an individual admits that its beliefs about the reality do not contain random variables and uncertain factors (the consequences of decisions depend on well-defined rules) [109].
15. **The principle of complementarity (inconsistency)** (N. Bohr, 1927; L. Zadeh, 1973). The high accuracy of description of a certain system is inconsistent with its high complexity [228]. Sometimes, this principle is given a simpler interpretation: the real complexity of a system and the accuracy of its description are roughly inversely proportional.
16. **The principle of monotonicity** ("keep the achieved," W. Ashby, 1952). In learning, self-organization, adaptation, etc., a system must "keep" an achieved (current) positive result (equilibrium, goal of learning, etc.) [14, 15].
17. **The principle of natural technologies** in biological systems (A. Ugolev, 1967 [205]). *The principle of block structure* states that physiological functions and their evolution are based on combinations of universal functional blocks implementing different elementary functions and operations.

At the first glance, the discussed principles of functioning of biological systems can be formally divided into natural-scientific approaches (e.g., principles no. 1, 2, 5, 8, and 15), empirical approaches (e.g., principles no. 4, 6, 10, 11, 14, 16, and 17) and intuitive approaches (principles no. 3, 7, 9, 12, and 13).

Natural-scientific approaches ("laws") reflect the general regularities, constraints and capabilities of biological systems imposed by natural laws. As a rule, empirical principles are formulated via analysis of experimental data, the results of experiments and observations, thereby having a more local character than natural-scientific approaches. And finally, intuitive laws and principles (in idea, not contradicting natural-scientific ones and being consistent with empirical ones) appear least formal and universal, as proceeding from intuitive understanding and common sense.

Yet, a detailed consideration shows that all the "natural-scientific" principles mentioned above are rather empirical and/or intuitive (not formally justified). For instance, the principle of least action (seemingly, a classical physical law) is formulated for mechanical systems (there exist its analogs in optics and other branches of physics). And its unadapted application to biological and other systems becomes somewhat incorrect and partially substantiated. In other words, that biological systems obey the principle of least action is merely a hypothesis made by researchers: today, in many cases it possesses no well-defined grounds.

Therefore, all the well-known and accepted principles (and laws) of biological systems functioning agree with one of the following standard statements: a regularity—"if a system has a (concrete) internal structure, then it demonstrates an (appropriate) behavior" or: a hypothesis—"if a system demonstrates a (concrete) behavior, then it most likely has an (appropriate) internal structure." Here the words "most likely" are essential: first-type statements establish sufficient conditions for realization of an observed behavior and can be (partially or completely) verified during experiments; second-type statements act as hypotheses, i.e., "necessary" conditions (in most cases, postulated without rigorous argumentation and bearing the explanatory function) which are imposed on the structure and properties of a system on the basis of its observation.

Particular laws and principles. We emphasize that different branches of control theory formulate separate laws and principles valid under corresponding assumptions. Here are some examples.

The book [59] presented several *laws of cybernetical physics*:

- The value of any controlled invariant of a free system can be changed in an arbitrary quantity via arbitrarily small feedback;
- For a controlled Lagrange or Hamilton system with a small dissipation rate ρ, the energy achievable by a control action of a level γ has the order of $(\gamma/\rho)^2$;
- Each controllable chaotic trajectory can be transformed into a periodic one using an arbitrarily small control action.

The book [157] introduced several *principles of control in organizations*:

- The *principle of agents' game decomposition*, stating that a *Principal* applies controls implementing a dominant strategy equilibrium of agents' game;
- The *principle of functioning periods' decomposition*, stating that a *Principal* applies controls making agents' decisions independent from game history;
- The principle of trust (*the fair play principle* [36, 39] and the *revelation principle* [141] as its analog), stating that an agent trusts information reported by a Principal, whereas the latter makes decisions assuming the truth of information reported by the former;
- The *principle of sufficient reflexion*, stating that the reflexion depth of an agent is defined by its awareness.

Obviously, the above and similar laws and principles represent fruitful and general results derived in separate branches of control theory, but have no universal character: they are inapplicable or selectively applicable in "adjacent" branches.

Control Principles[8] [152]

Principle 1 (*the principle of hierarchy*). Generally, a control system has a hierarchical structure. It must agree with the functional structure of a controlled system

---

[8]Of course, ideally all principles should not be stated as requirements to control systems ("it must be that…", "it is necessary that…" and so on), which can satisfied or not satisfied. Instead, the general approach should be, whenever a certain principle fails, a control system appears unable to work properly. Unfortunately, such "hard" principles do not exist (perhaps, except the feasibility of control).

and not contradict the hierarchy of (horizontally or vertically) adjacent systems. Tasks and resources supporting the activity of a controlled system must be decomposed according to its structure.

Principle 2 (*the principle of unification*). Controlled systems and control systems of all levels must be described and studied using common principles (this applies both to the parameters of their models and the efficiency criteria of their functioning). However, such principles must not eliminate the necessity of considering the specifics of a concrete system. Most real control situations can be reduced to a set of the so-called typical situations, where the corresponding *typical decisions* appear optimal.

On the other part, control inevitably causes *specialization* (restriction of variety) of control subjects and controlled subjects.

Principle 3 (*the principle of purposefulness*). Any impact of a control system on a controlled system must be purposeful.

Principle 4 (*the principle of openness*). Operation of a control system must be open to information, innovations, etc.

Principle 5 (*the principle of efficiency*). A control system must implement the most efficient control actions from the set of feasible control actions (also see the principle of extremization).

Principle 6 (*the principle of responsibility*). A control system appears responsible for decisions made and the efficiency of controlled system operation.

Principle 7 (*the principle of non-interference*). Any-level Principal interferes in a process iff its direct subordinates are unable to implement a complex of necessary functions (at present and/or based on a forecast).

Principle 8 (*the principle of social and state control, participation*). Control of a social system must aim at the maximal involvement of all interested subjects (society, bodies of state power, individual and artificial persons) in the development of a controlled system and its operation.

Principle 9 (*the principle of development*). A control action lies in modifying a control system proper (being induced from within, it can be treated as self-development). The matter also concerns the development of a controlled system.

Principle 10 (*the principle of completeness and prediction*). Under a given range of external conditions, the set of control actions must ensure posed goals (the completeness requirement) in an optimal and/or feasible way. This must be done taking into account a possible response of a controlled system to certain control actions in predicted external conditions.

Principle 11 (*the principle of regulation and resource provision*). Control activity must be regulated (standardized) and correspond to constraints set by a metasystem (a system possessing a higher hierarchical level). Any management decision or control action must be feasible (also, in the sense of provision with necessary resources).

Principle 12 (*the principle of feedback*). Efficient control generally requires information on the state of a controlled system and on the conditions of its functioning. Moreover, implementation of a control action and corresponding consequences must be monitored by a control subject.

Principle 13 (*the principle of adequacy*). A control system (its structure, complexity, functions) must be adequate to a controlled system (to its structure, complexity, functions, respectively). Problems to-be-solved by a controlled system must be adequate to its capabilities.

Principle 14 (*the principle of well-timed control*). This principle states that, in real-time control, information required for decision-making must be supplied at the right time. Moreover, management decisions (control actions) must be made and implemented (chosen and generated, respectively) quickly enough according to any changes in a controlled system and external conditions of its functioning. In other words, the characteristic time of management decisions or control actions must not exceed the characteristic time of changes in a controlled system (i.e., a control system must be adequate to controlled processes in the sense of their rate of change).

Principle 15 (*principle of predictive reflection*). A complex adaptive system predicts feasible changes in essential external parameters. Consequently, when generating control actions, one should predict and anticipate such changes.

Principle 16 (*the principle of adaptivity*). The principle of predictive reflection underlines the necessity of predicting the states of a controlled system and corresponding actions of a Principal. In contrast, the principle of adaptivity states that (1) one must consider all available information on the history of controlled system functioning and (2) once made decisions or chosen control actions (and the corresponding principles of decision-making) must be regularly revised (see the principle of well-timed control) following any changes in the states of a controlled system and in the conditions of its functioning.

Principle 17 (*the principle of rational decentralization*). This principle claims that, in any complex multi-level system, there exists a rational decentralization level for control, authorities, responsibility, awareness, resources, etc. Rational decentralization implies adequate decomposition and aggregation of goals, problems, functions, resources, and so on.

In [154] it was shown that *multilevel hierarchical systems* gain new properties (in comparison with two-level ones) mainly due to the following factors:

– the *"aggregative" factor*, consisting in aggregation ("convolution," "compression," and so on) of information about system elements, subsystems, an environment, etc. as the level of hierarchy grows;
– the *"economic" factor*, consisting in variation of financial, material and other resources of a system under any changes in the composition of its components;
– the *"uncertainty" factor*, consisting in variation of the awareness of system elements about the essential (internal and external) parameters of their functioning;

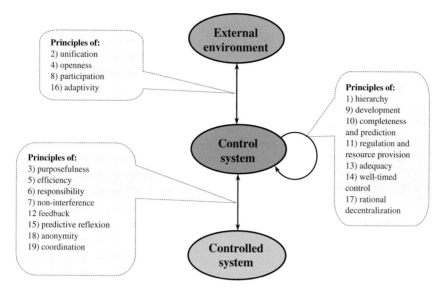

Fig. 3.2 Control principles: A classification based on the relations between objects

– *the "organizational" factor*, consisting in power sharing, i.e., the feasibility of some system elements to establish "rules of play" for the other;
– *the "informational" factor*, consisting in variation of informational load on system elements.

"In fact, any complex system, whether it has arisen naturally or been created by human beings, can be considered organized only if it is based on some kind of hierarchy or interweaving of several hierarchies. At least we do not yet know any organized systems that are arranged differently." [203, p. 37].

Principle 18 (*the principle of democratic control*, also known as *the principle of anonymity*). This principle requires equal conditions and opportunities for all participants of a controlled system (without a priori discrimination in informational, material, financial, educational and other resources).

Principle 19 (*the principle of coordination*). This principle declares that, under existing institutional constraints, control actions must be maximally coordinated with the interests and preferences of controlled subjects.

Principle 20 (*the principle of ethics, the principle of humanism*) implies that, in management and control, consideration of existing ethical norms (in a society or an organization) has a higher priority over other criteria.

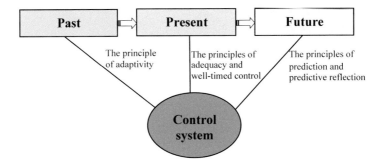

Fig. 3.3 Control principles: A classification based on the temporal relations

Note that the above control principles are applicable almost to any-nature systems (probably, except the principle of social and state control and the principle of coordination, making no sense in control of technical systems).

Possible classification bases for the listed control principles are the relations between objects (a controlled system, a control system, an external environment—see Fig. 3.2) or the temporal relations (past, present, future—see Fig. 3.3).

Therefore, the general laws and principles of control are the subject of Cybernetics. Their list is far from final canonization, and its supplementation and systematization represent a major task of Cybernetics!

Chapter 4
Systems Theory and Systems Analysis. Systems Engineering

Logically and historically, the content of cybernetics has indissoluble connection with the category of "*system*" (see Appendix A). Here the key role belongs to two terms—systems approach and systems analysis.

From the historical perspective, general systems analysis appeared within the framework of *general systems theory* (GST) founded by biologist L. Bertalanffy: in the 1930s he proposed the concept of an *open system* [27]. The first complex publications on GST were [25, 26], see also [30, 163, 177]. Interestingly, the term "systems analysis" originated in RAND Corporation reports dating back to 1948 (the first book was [92]).

The later development of systems analysis in the USSR (Russia) and other countries was different. First of all, systems analysis was assigned nonidentical interpretations. Our discussion begins with the traditions of the Russian scientific schools.

SYSTEMS APPROACH is a direction in the methodology of scientific cognition and social practice, which studies objects as **systems**, i.e., an integral[1] set of *elements* in the aggregate of their *relations* and *connections*.[2] Systems approach facilitates adequate problem formulation in concrete sciences and gives efficient strategies of their study.

Systems approach is a general way of activity organization, which embraces any type of activity, reveals regularities and interconnections for their efficient usage [148].

SYSTEMS ANALYSIS ("a practical methodology of problem solving") is a set of methods oriented towards analysis of complex *systems* (technical, economic, ecological, educational and other ones).

As a rule, *systems studies* result in a *choice* of a well-defined alternative (a development program of an organization or a region, design parameters, etc.). *Systems approach* is valuable, since consideration of systems analysis categories underlies general logical and sequential solution of control and decision-making problems.

---

[1]Integrity and commitment to a common goal form a backbone factor.
[2]An aggregate of stable connections among system elements, ensuring its integrity and self-identity, is called its structure.

© Springer International Publishing Switzerland 2016
D.A. Novikov, *Cybernetics*, Studies in Systems, Decision and Control 47,
DOI 10.1007/978-3-319-27397-6_4

The efficiency of problem solving using systems analysis depends on the structure of problems [148].

Being remarkable for its interdisciplinary status, systems analysis considers, e.g., an activity as a complex system aiming at elaboration, substantiation and implementation of complex problem solving including political, social, economic, technical and other problems [166].

To solve *well-defined problems* (i.e., the ones which admit an explicit quantitative description and strong formalization), systems analysis employs optimization and operations research methods: a researcher constructs an adequate mathematical model and seeks for optimal purposeful actions (control) within the model. To solve *ill-defined problems*, systems analysis operates different techniques including typical stages (see Table 4.1 for a series of common approaches to systems analysis and strategic analysis of problem solving). Actually, systems analysis suggests universal methods of problem solving applicable to a wide range of fields: organizational control, economics, military science, engineering, and others.

Therefore, in the USSR systems analysis was considered side by side with systems theory (and later almost "absorbed" the latter) as a set of general principles of examining any systems (systems approach). Similarly to cybernetics, systems analysis (being an integrative science) admits the "umbrella" definition as a union of different component sciences under the auspices of "systemacy": artificial intelligence, operations research,[3] decision theory, systems engineering and others, see Fig. 4.1. According to this viewpoint, systems analysis has almost no its own results.

This result has definite causes: historically, systems analysis appeared via development of *operations research* (see the first book [139], the classical textbook [217],[4] the modern textbooks [82, 198]) and *systems engineering* (see the first book [70]). With the course of time, *operations research* transformed into *management science*[5] with basic applications to control of organizational and production systems [17]. Nowadays, many Russian scientists (e.g., [110, 136, 182, 197]) still understand systems analysis as an aggregate of methods of optimization, operations research, decision-making, mathematical statistics and others, in addition to the concept of systemacy proper. In this context, we refer to the classical textbook [166]. For an interested reader, the publications [197, 213] survey the history of systems analysis development in the USSR and Russia.

The second interpretation of systems analysis (by analogy with Cybernetics, *Systems analysis* with capital S—compare Figs. 1.9 and 4.1) covers the general laws, regularities, principles, etc. of functioning and exploration of different-nature systems.

---

[3]Systems analysis and operations research are correlated as strategy and tactics, see [92, p. 1].

[4]The classical range of operations research includes choice problems, multicriteria decision-making, linear, nonlinear and dynamic programming, Markov processes, queuing theory, game-theoretic methods in decision-making, networked planning and reliability theory.

[5]S. Beer defined management science as "the business use of operations research.".

Table 4.1 Systems analysis and strategic analysis of problem solving (see [73])

E. Golubkov	P. Drucker	D. Novikov	S. Optner	N. Fedorenko	Yu. Chernyak	S. Young
1. Problem statement	1. Purpose and expected results	1. Monitoring and analysis of actual state	1. Symptoms identification	1. Problem formulation	1. Problem analysis	1. Goal-setting for organization
2. Examination	2. Key elements to process design: time, resources, budget, major steps	2. Forecasting of evolution	2. Problem urgency estimation	2. Definition of goals	2. Definition of system	2. Problem identification
3. Analysis	3. Roles and responsibilities of self-assessment team	3. Goal-setting	3. Goal-setting	3. Data acquisition	3. Structural analysis	3. Diagnosing
4. Preliminary judgment	4. Elements essential to success:	4. Choosing technology of activity	4. Definition of system structure and its defects	4. Elaboration of the maximal number of alternatives	4. Formation of general goal and criterion	4. Decision search
5. Confirmation	– Utilizing an experienced facilitator;	5. Planning and resources allocation	5. Capabilities assessment	5. Selection of alternatives	5. Goal decomposition, identification of demands in resources and processes	5. Assessment and choice of alternatives
6. Final judgment	– Engaging dispersed leadership;	6. Motivation	6. Alternatives search	6. Modeling by equations, programs or scenarios	6. Identification of resources and processes	6. Decision negotiation
7. Implementation of chosen decision	– Encouraging constructive dissent;	7. Control and operative management	7. Alternatives assessment	7. Costs estimation	7. Forecasting and analysis of future conditions	7. Decision approval
	– Using data to inform dialogue	8. Reflexion, analysis and improvement of activity	8. Decision elaboration	8. Sensitivity tests (parametric analysis)	8. Assessment of goals and means	8. Preparation for decision implementation
			9. Decision acceptance		9. Selection of alternatives	9. Decision application
			10. Decision procedure initiation		10. Diagnosis of existing system	10. Efficiency control
			11. Decision implementation control		11. Elaboration of complex development program	verification
			12. Assessment of implemented decision and its consequences		12. Design of organization for goals' achievement	

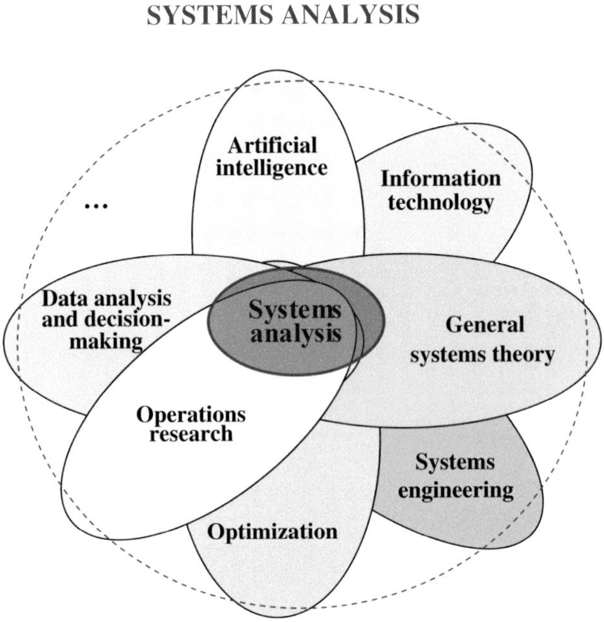

SYSTEMS ANALYSIS

Fig. 4.1 The composition and structure of systems analysis

Here the main body of scientific results is the philosophical and conceptual aspects of systems analysis and general systems theory, see [28, 46, 184, 204].

Among Soviet and Russian scientific schools focused on Systems analysis, we emphasize two fruitful theoretical and applied research groups, viz., *the methodological school* of Schedrovitsky [186] and the followers of S. Nikanorov—"the school of *conceptual analysis and design* of organizational control systems" [146]. The both schools operate the categories of system, control, organization and methodology, as well as seek to analyze and synthesize most general solution methods for a wide range of problems. In other words, they are inseparably linked with Cybernetics.

Systems analysis, just like cybernetics, endures the "romantic" period and the period of disillusions (see Sect. 1.3). "Presently, the terms "analysis of systems" or "systems analysis" often excite the antithetical feelings of different people. On the one part, here is faith in the omnipotence of the new approach capable of solving difficult and large-scale problems and, on the other part, charges of dalliance decorated by a fashionable terminology." [114]. These words of O. Larichev preserve their topicality even now. Both Cybernetics and Systems Analysis need GENERAL results including generalizations from intensively developing sciences in the "umbrella brand" of systems analysis (see Fig. 4.1).

Systems theory and systems engineering. Let us analyze "systems" terminology in the English segment of publications. The high level of abstracting and generality of systems studies in the USSR and Russia corresponds to the English terms "general systems theory" (initially) and *systems science* (present days). In other words, "systems analysis" as it is comprehended in Russia rather matches "systems science" (SS) in foreign research, as sciences about systems, systems studies—see Fig. 4.2.

As a matter of fact, general systems science evolved in several directions. First, its "mainstream" gave birth to two known subdirections: K. Boulding's *theory of systems classes* [30] and P. Checkland's *soft systems methodology* [43, 44].

Second, note that the 1950–1970s were remarkable for a significant break-through in *mathematical system theory* [40, 93, 132, 133], which later merged with control theory.

Third, we naturally mention *systems dynamics* exploring the influence of system elements and structure on its behavior in time. Here the main apparatus includes simulation modeling of differential equations or discrete mappings. The pioneering works were [57, 58] and the most famous application to global development was described in the book [130]. The state-of-the-art in this field can be traced in [78, 129].

Reverting to the subject, *systems analysis* actually concerns any analytical study assisting a decision-maker to choose an appropriate course of actions [147].

Subsequently, SA developed towards *systems engineering* (SE) (see the classical publications [67, 70]). This is a branch of science and technology covering the whole *life cycle* of a complex system (design, production, testing, exploitation, support, maintenance and repair, upgrade and utilization).

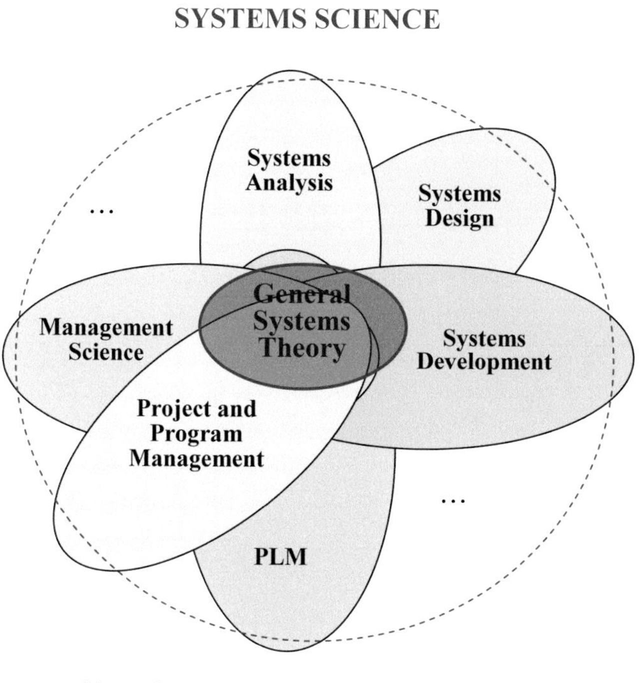

SYSTEMS SCIENCE

Fig. 4.2 The composition and structure of systems science

As years passed, SA became a set of practice-oriented analysis technologies for concrete systems, i.e., products and/or services [142, 185, 209, 219]. Systems analysis goes in parallel with systems design (SD), systems development and other associated stages.

Nowadays, SS and SE (e.g., see the modern textbooks and standards [49, 76, 88, 185, 196, 209, 219]) comprise SA, SD, *product lifecycle management* (PLM), *project and program management*, several branches of management science and others, as illustrated by Fig. 4.2. And general systems theory forms their *common methodological core*, see Fig. 4.2.

Most applications of SE are complex technical and organization-technical systems, as well as software development.

Presently, the branches of "systems studies" (SA, SD, …) are rather an aggregate of *technologies* and a common *language* in the form of *standards* (arising through generalization of *successful practical experience*, see the Conclusion) than scientific directions.

Systems of systems. An intensively developing direction of systems theory and systems engineering embraces the problematique of a so-called *system of systems*; it considers the interaction of autonomous (self-sufficient) systems jointly forming an integral system with its own goals, functions, etc. Among examples, we refer to networks of networks, SmartGrid in power engineering, the interaction of units and corps in military science, complex production processes, and so on. This direction employs the concept of *holism*[6] [190] and dates back to the late 1960s (see the classical paper [1] by R. Ackoff). A good survey of latest achievements can be found in [90].

---

[6]Holism is an approach treating complex systems as a whole; it claims that the properties of complex systems cannot be derived via examining the properties of their elements.

Chapter 5
Some Trends and Forecasts

Any mature science necessarily predicts its own development and the development of adjacent sciences. As Cybernetics represents a *metascience* (see Chap. 2) with respect to its components–control theory and others, its functions should include *analyzing* their *trends*, seeking for generalizations and *forecasting*. Ideally, the matter concerns normative forecasting, i.e., constructing a multi-alternative scenario forecast with separation of desired trajectories and an action plan for their implementation.

The current chapter reveals a series of trends in control theory (their list does not claim to be exhaustive, rather a call for such activities). Particularly, we consider in brief the topic structure of some leading control conferences (Sect. 5.1), interdisciplinarity (Sect. 5.2), "networkism" (Sect. 5.3), heterogeneous models and hierarchical modeling (Sect. 5.4), "intellectualization" and reflexion (Sect. 5.5), big data and big control (Sect. 5.6). However, our discussion does not touch internal paradigm problems of different branches in control theory including the effects of their "linear" development, aspiration towards self-isolation[1] and others.

5.1 Topic Analysis of Leading Control Conferences

Nowadays, several hundreds of scientific conferences (seminars, symposia, meetings, etc.) are organized yearly worldwide on certain aspects of control theory and applications. Yet, there are a few "emblematic" leading scientific events reflecting and predetermining basic trends [150]. Being subjective and not pretending to a complete overview, the author emphasizes triennial world congresses conducted by *International Federation of Automatic Control* (IFAC) and annual *Conferences on Decisions and Control* (CDC) under the auspices of *Institute of Electrical and Electronics Engineers* (IEEE). Alongside with these major events (or even jointly with CDC), there exist regular "national"[2] conferences: *American Control*

---

[1]The existing grant-based funding of research facilitates differentiation of sciences and partially stimulates the existence of scientific self-reproducing "sects" in all fields of investigations.
[2]Actually, these conferences gather researchers from many other countries.

© Springer International Publishing Switzerland 2016
D.A. Novikov, *Cybernetics*, Studies in Systems, Decision and Control 47,
DOI 10.1007/978-3-319-27397-6_5

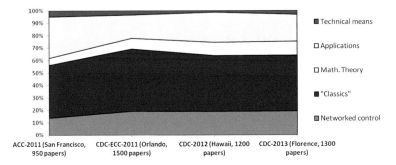

Fig. 5.1 The general topics of ACC and CDC

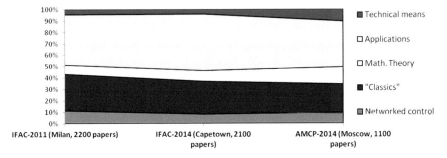

Fig. 5.2 The general topics of IFAC congresses and AMCP-2014

Conference (ACC) and *European Control Conference* (ECC). In the USSR, the role of such national conferences belonged to All-Union Meetings on Regulation Theory (later, on Automatic Control and, then, on Control Problems). Interestingly, the gradually changing title of these scientific events agrees with the evolution of control theory and its subjects (see below).

Generally speaking, world science demonstrates a stable growth of publications dedicated to control (see Figs. 1.4–1.6).

Figures 5.1, 5.2 and 5.3 present the "quantitative" comparison[3] of the topics at 2011 and 2014 IFAC World Congresses (also, see [171]), ACC-2011, CDC-ECC-2011, CDC-2012, CDC-2013 and *All-Russia Meeting on Control Problems* (AMCP-2014).[4]

General topics. The author has classified the papers, being mostly concerned with relative (not absolute) indexes: they reflect the current distribution and dynamics of priorities, despite the subjectiveness and certain arbitrariness of

---

[3]All figures show the relative shares of papers having a corresponding topic.

[4]In AMCP-2014, about 25–33 % of the papers were dedicated to control problems in interdisciplinary systems (socioeconomic, organizational and technical, etc.). They have been eliminated from our analysis.

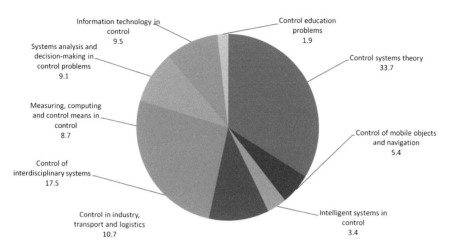

Fig. 5.3 The topics of papers at AMCP-2014

classification bases. The following groups of topics have been identified via expertise: *mathematical control theory* (mathematical results invariant with respect to application domains of controlled objects), "*classics*" (automatic control theory (ACT) in a wide interpretation[5]), "*networked control*" (covering situations when a control object and/or subject and/or communication between them has a networked structure), *technical means of control, applied control problems*, see Figs. 5.1, 5.2 and 5.3.

Figures 5.1 and 5.2 illustrate (a) the relative "*stability of traditions*" of appropriate scientific events and (b) a well-known fact that CDC are more "theoretical," whereas IFAC congresses are par excellence application-oriented. In this sense, AMCP-2014 most likely follows the tradition of IFAC congresses.

Networked control. We emphasize the growing interest of researchers in *networked control* problems (the number of papers in peer-reviewed journals has almost doubled within 5–6 years). This observation also follows from the analysis of publications indexed by Web of Science, see Fig. 5.4.

Figures 5.5 and 5.6 specify the topics of networked control by the levels of agents architecture in *multi-agent systems* (MAS) and problems treated at these levels (see Sect. 5.3). The following groups of topics have been identified via expertise: MAS and *consensus* problems, communications in MAS, *cooperative control*, upper levels of control (*strategic behavior* of agents), "others" (mostly, information and communication networks with a slight emphasis on control problems).

---

[5]Notwithstanding its "classical" character, ACT has an intensive development, including the appearance of new problems in well-known fields (e.g., in linear control systems) and new controlled objects (e.g., the rapid growth of publications on quantum systems control).

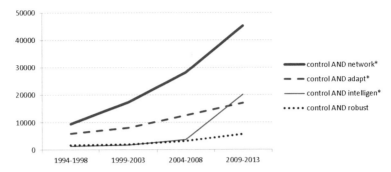

Fig. 5.4 The number of papers on networked control published worldwide (according to Web of Science)

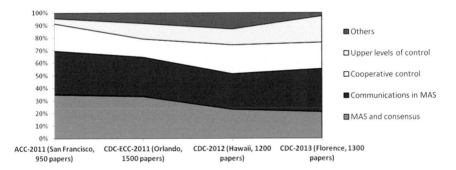

Fig. 5.5 Specification of networked control topics at ACC and CDC

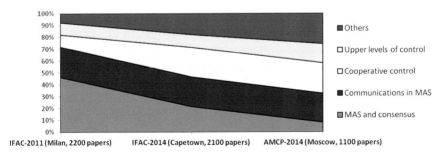

Fig. 5.6 Specification of networked control topics at IFAC congresses and AMCP-2014

According to Figs. 5.5 and 5.6, investigators gradually shift their efforts towards higher levels of agents' architecture, i.e., from consensus and communications problems to cooperative control and strategic behavior models of agents [56].

Applications. Figures 5.7 and 5.8 specify the internal structure of applied topics at the scientific events under consideration. The following groups of applications

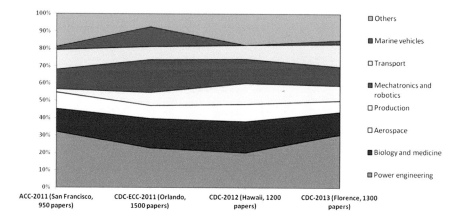

Fig. 5.7 Specification of applied topics at ACC and CDC

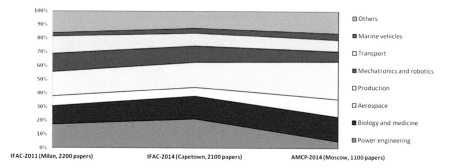

Fig. 5.8 Specification of applied topics at IFAC congresses and AMCP-2014

have been identified via expertise: *power engineering, biology and medicine, aerospace, production* (mostly, industrial production), *mechatronics* and robotics, *transport* (mostly, automobile transport and traffic), *marine vehicles* and "others" (from agriculture to education).

Clearly, in recent years an emphasis has been gradually changing from traditional control problems in production and telecommunication systems to power engineering and biomedical applications.

5.2 Interdisciplinarity

Modern control theory (see Figs. 5.9 and 5.11) studies control problems for different classes of controlled *objects* by designing or applying appropriate *methods* and *means* of control.

Fig. 5.9 Controlled objects, methods and means of control

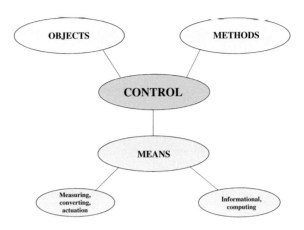

Fig. 5.10 The life cycle of control theory for a certain class of controlled objects

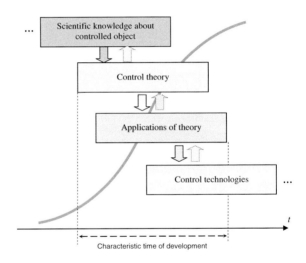

The term *"interdisciplinarity"* as staying at the junction of sciences,[6] their branches, etc. reflects the variety of controlled objects and the variety of methods and means of control. (Interdisciplinarity with capital I reflects their generality). This subsection mostly deals with the variety of controlled objects.

For a certain class of controlled objects, *the life cycle structure of control theory* is illustrated by Fig. 5.10. Using information acquired by a corresponding science about a controlled object,[7] control experts formulate appropriate models and

---

[6] According to Merriam-Webster dictionary, science is knowledge about or study of the natural world based on facts learned through experiments and observation; a particular area of scientific study (such as biology, physics, or chemistry) or a particular branch of science; a subject that is formally studied in a college, university, etc.

[7] In the case of technical systems, initial information "suppliers" are mechanics, aerodynamics, and so on.

perform their theoretical study (analysis and synthesis of control actions, exploration of different properties such as observability, identifiability, controllability, stability and others).

Subsequently, the theory finds applications implemented in the form of control technologies. From the temporal viewpoint, each class of controlled objects perhaps demonstrates its own characteristic time of development and "golden period" with the maximum pace of results accumulation (see Fig. 5.10).

Control theory embraced various controlled subjects and objects during more than a century and a half of its development, see Fig. 5.11. In the area of *technical* and *organization-technical systems*, the main emphasis has been recently shifted to decentralized intelligent systems (see Sect. 5.5). For instance, more and more research works are dedicated to *upper control levels* in the terminology of their types' hierarchy [211]:

1. programmed control;
2. feedback control;
3. robust control;
4. adaptive control;
5. intelligent control;
6. intellectual (smart) control (in contrast to intelligence, intellectuality means the presence of autonomous goal-setting (autonomous and adaptive generation of efficiency criteria) in control loops).

Within the last 50 years,[8] mathematical control theory simultaneously involved new and new classes of controlled objects (since the 1950–1960s—*economic systems*, later *ecological-economic* and other *systems*). Concerning the recent decades, the focus of attention has been gradually drifting to *living systems* and *social systems*. The fruitful development of the corresponding branches of control theory and accumulation of knowledge about controlled objects require a close cooperation between mathematicians (control experts) and representatives of associated sciences.

Moreover, the application domain of control theory becomes wider. A key problem in its methods dissemination (*the integration problem*) is the availability of sufficiently adequate models of controlled objects. Again, here we need a close cooperation between control experts and representatives of associated sciences (physics, economics, biology, sociology and others).

---

[8]Interestingly, a broader retrospective review indicates that social systems cyclically interchange with technical ones in the focus of control theory, getting "back" at a new turn of the dialectical spiral. Indeed, perhaps the first object of control (in the prehistoric society) was a group of people, later on—transport and elementary mechanisms, again followed by groups of people (Plato-N. Machiavelli-F. Bacon-T. Gobbs-...-A. Ampere-B. Trentowski). Starting from the middle of the 19th century, control theory switched to technical (mechanical) systems. Today, control of human beings, their groups and/or collectives is again on the agenda.

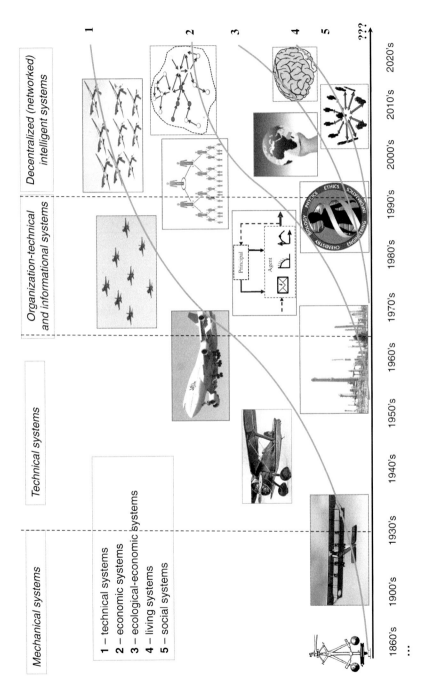

Fig. 5.11 The past, present and future of control theory

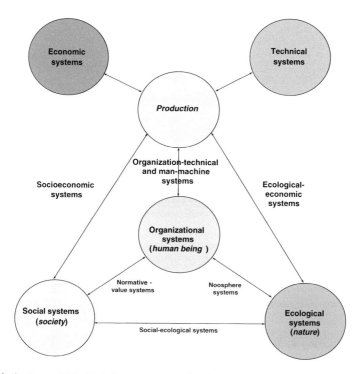

Fig. 5.12 Systems of interdisciplinary nature: a classification

For a large scientific organization, institution or scientific school to maintain and/or gain leading positions in the field of control in several decades when seeming new objects of control will become classical, it is necessary to initiate their intensive research right now!

More frequently, controlled objects represent the so-called *interdisciplinary-nature systems* [152]. Imagine that the corresponding classification is based on the subject of human activity ("nature—society—production"). In this case, we may distinguish among organizational systems (people), ecological systems (nature), social systems (society), as well as economic (technical) systems (production), see Fig. 5.12. Different paired combinations emerge at the junction of these classes of systems:

- organization-technical systems;
- socio-economic systems;
- ecological-economic systems;
- socio-ecological systems;

- normative-value systems;
- noosphere systems.[9]

Separation of the following *research priorities* evidences the growing interest of investigators in interdisciplinary-nature systems:

- US National Science Foundation: group control, spacecraft clusters, combat control, control of financial and economic systems, control of biological and ecological systems, multiple-profile teams in control loop, etc.;
- Research in the European Union: man-machine symbiosis (modeling of a human being in control loops including the case of a controlled subject), complex distributed systems and quality improvement of systems in an uncertain environment (global manufacturing, security, heterogeneous control strategies, new principles of multidisciplinary coordination and control) and others;
- Key directions of fundamental research by the Russian Academy of Sciences: control in interdisciplinary models of organizational, social, economic, biological and ecological systems; group control; cooperative control and others.

The paper [66] mentioned three global challenges to cybernetics, namely, transitions:

1. from nonliving to living (from chemistry to biology);
2. from living to intelligent (from living organisms to human consciousness);
3. from human consciousness to human spirit as the highest level of consciousness.

The specifics of interdisciplinary-nature systems incorporating human beings as a control object consist in the following:

- *independent goal-setting, purposeful behavior* (conscious information misrepresentation and strategic behavior, non-fulfillment of commitments, etc.);
- *reflexion* (nontrivial mutual awareness, foresight, behavior forecasting for a Principal or control object/subject, the effect of roles exchange,[10] etc.);
- *bounded rationality* (decision-making in uncertain conditions and under existing constraints on the volume of processed data);
- *cooperative* and/or *competitive interaction* (formation of coalitions, informational contagion, etc.);
- *hierarchical structure*;
- *multicomponent structure*;
- *distributed/networked structure* and/or *different scale* (in space and/or time, see the paper [135] discussing the principle of requisite variety and its extension to multiscale systems).

---

[9]Systems, where a specially organized activity of human beings is a determining factor for the development of large-scale (global) ecological systems.

[10]In systems whose elements have strategic behavior, discrimination between control subjects and controlled ones can be ambiguous; e.g., in some situations a subordinate manipulates its superior.

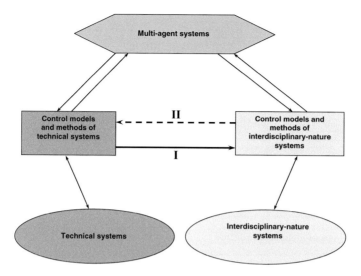

Fig. 5.13 Results transfer

Historically, "mechanical" systems (later, technical ones) were the first classes of controlled objects theoretically studied on a mass scale (see Fig. 5.11). As a matter of fact, most deep and extensive theoretical results of control were obtained exactly for these classes. As new controlled objects appear, researchers naturally endeavor to perform "*results transfer*," i.e., translate some existing results to the new objects. That was exactly the case for interdisciplinary-nature systems: general results of Cybernetics and concrete analysis results of control problems for technical systems were transferred to the former, see arrow I in Fig. 5.13.

Following accumulation of its own results within the framework of control models and methods of "non-mechanical" (e.g., living systems[11]) and/or interdisciplinary-nature systems (e.g., socioeconomic systems), the inverse tendency has been gradually showing itself—more and more artificial technical or informational systems are assigned the inherent properties of social or living systems. This represents a basic trend which will be perhaps intensified in future. In many cases, *multi-agent systems* act as a tool of "*inverse results translation*" (see arrow II in Fig. 5.13). Multi-agent systems are discussed below. For instance, such inverse translation takes place in numerous manifestations of "intellectuality": cooperative behavior, reflexion, etc.

---

[11]For the sake of justice, note that at all times living systems encouraged scientists and engineers to apply analogies, i.e., to "repeat" certain properties of living nature objects in artificial systems.

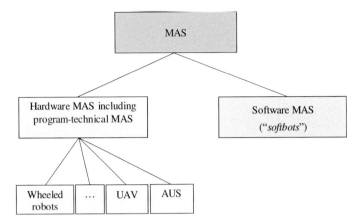

Fig. 5.14 Types of multi-agent systems

5.3 "Networkism"

For the recent 15 years, a modern tendency in control theory has been seeking towards "*miniaturization*,"[12] "*decentralization*" and "*intellectualization*" in systems of very many interacting autonomous *agents* having social, technical or informational nature. Inherent properties of multi-agent systems (MAS) such as *decentralized* interaction and agents' multiplicity induce fundamentally new emergent properties (*autonomy*, smaller *vulnerability* to unfavorable factors, etc.) crucial in several applications [180, 189, 226].

MAS can be divided into *hardware* (pioneering publications dating back to the middle of the 1990s) and *software* ones (since the middle of the 1970s), as illustrated by Fig. 5.14. The former include mobile robots (wheeled robots, unmanned aerial vehicles (UAV), autonomous unmanned submersibles (AUS), etc.), control systems of complex industrial and technological objects (computer-aided control systems of industrial processes, power engineering–SmartGrid and so on). The latter include control systems, where agents are *softbots*, i.e., autonomous programmed modules solving *distributed optimization* problems according to established protocols (possible applications are logistics systems in manufacturing and transport, softbots in digital networks, i.e., real-time scheduling, assignment of functions and tasks, and so on).

On the other hand, a striking tendency of recent 10–15 years concerns transition from *centralized control* (a same control system responsible for each of several controlled objects, e.g., agents including their pairwise interactions) to

---

[12]Control problems of quantum systems are mostly treated in theory, but micro-level controlled objects ("microsystems") have become almost common.

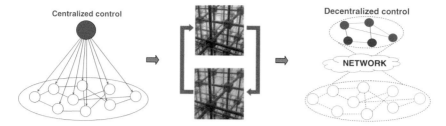

Fig. 5.15 Decentralized control

decentralized control (a control network is superstructed over a *network* of interacting objects), and then to *communication* between control systems and agents via a network. Here a separate problem lies in control of this network, see Fig. 5.15. Networked MAS are considered in the next section.

Consequently, today "*networkism*" exists in controlled objects, control systems and their interaction. In many cases, a control system is even "*immersed*" into a controlled object, thereby forming an integrated (perhaps, hierarchically organized) network of interacting agents. The number of research works on networks (in the wider interpretation,[13] information and communication technologies (ICT), Internet and other technologies in complex distributed systems) is huge and still continues to grow (see Sect. 5.1).

Today, the overwhelming majority of multi-agent systems investigations are theoretical, despite their mass character. As a rule, consideration gets confined to computing experiments, and there exist merely a small number of open-access publications describing real applications of MAS.

The forthcoming years will be remarkable for transition from the so-called C^3 *paradigm* (joint solution of *Control + Computations + Communications* problems) to *the C^5 concept* (*Control + Computations + Communications + Costs + Life Cycle*). Here the above-mentioned problems are solved taking into account cost aspects (in the general sense) over the whole *life cycle* of a system including the joint design of a control system and its controlled object.

Speaking about "*networkism*," we have to touch "*network-centrism*"[14] extremely fashionable nowadays (also called "*network-centric fever*"). It admits several interpretations covering organization and analysis principles of any networks in principle or temporary networks created for specific task or mission execution at a right place and right time (*networked organizations*, e.g., interaction of military

---

[13]Not to mention the penetration of ICT into engineering and everyday life, the associated educatory and social capabilities and threats.

[14]Network-centrism operates its own abbreviations differing from control theory (see above): C^3I—Command, Control, Communications and Intelligence, C^4I—Command, Control, Communications, Computers and Intelligence, and others.

units in a combat theater). This approach finds wide application in network-centric warfare problems for vertical and horizontal integration of all elements during a military operation (control, communication, reconnaissance and annihilation systems).

Another manifestation of "*networkism*" concerns the growing popularity of *distributed decision support systems*. The intensive development of ICT increases the role of informational aspects of control in decentralized hierarchical systems (an example is decision-making support in distributed decision systems which integrate heterogeneous information on strategic planning and forecasting from different government authorities and industrial sectors). One of such aspects consists in *informational control* as a purposeful impact on the awareness of controlled subjects; therefore, a topical problem is to develop a mathematical apparatus providing an adequate description for an existing relationship between the behavior of system participants and their mutual awareness [158].

Design of intelligent analytic systems for informational and analytic support of goal-setting and control cycle represents another important informational aspect of control in decentralized hierarchical systems. Here it seems necessary to substantiate methodological approaches to control efficiency in decentralized control systems, including elaboration of principles and intelligent technologies for data acquisition, representation, storage and exchange.

We underline that an appreciable share of information required for situation assessment, goal-setting and control strategy choice in decentralized systems is ill-structured (mostly, in the form of text). And there arise the problems of relevant search and further analysis of such information. The described circumstances bring to the need for suggesting new information retrieval methods (or even knowledge processing methods) based on proper consideration of its lexis and different quantitative characteristics and, moreover, on analysis of its semantics, separation of target data and situation parameters, assessment of their dynamics and scenario modeling of situation development in future periods.

5.4 Heterogeneous Models and Hierarchical Modeling

In recent years, control theory more and more addresses the term of system "*heterogeneity*" comprehended, in the first place, as the multiplicity of its mathematical description (e.g., descriptive dissimilarity of separate subsystems: the type and scale of time/space of subsystems functioning, multi-type descriptive languages for certain regularities of a studied object, etc.). "Heterogeneity" also means complexity appearing in (qualitative, temporal and functional) *dissimilarity*, (spatial and temporal) *distribution* and the *hierarchical/networked structure* of a controlled object and an associated control system (see Sect. 5.3).

An adequate technology for design and joint analysis of a certain set of heterogeneous systems models is the so-called *hierarchical modeling*. According to this technology, models describing different parts of a studied system or its different

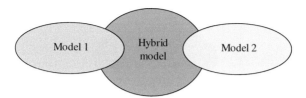

Fig. 5.16 The narrow interpretation of a hybrid model

properties (perhaps, with different levels of detail) are ordered on the basis of some logic, thereby forming a hierarchy or a sequence (a horizontal chain). Generally, lower hierarchical levels correspond to higher levels of detail in modeled systems description. Each element of a sequence possesses almost same level of detail, and the results (outputs) of a current model represent input data for a next model. Such approach to modeling was born and further developed in the 1960–1970s [40, 133].

In some sense, hierarchical models are a wider category than hybrid models and the multi-model approach. A *hybrid model* is a model combining elements of two or more models reflecting different aspects of a studied phenomenon or process and/or employing different apparatuses (languages) of modeling–see Fig. 5.16. For instance, a hybrid model can include discrete and continuous submodels, digital and analog submodels, and so on.

In the wider interpretation, a hybrid model represents a complex of models each chosen under well-defined conditions, see Fig. 5.17. As an example, consider hybrid dynamic systems (HDS, also known as switching systems). The expression in the right-hand side of the HDS differential equation is chosen from a given set of options depending on the current state of the system and/or time and/or auxiliary conditions.

Within the *multi-model approach*, several models are used sequentially or simultaneously with further or current analysis and selection of "best" results.

Hierarchical (sequential) models may have a more complex structure, see Fig. 5.18. At each level, a model can be hybrid or follow the multi-model approach. Hierarchical models lead to the problems of *aggregation* and *decomposition* well-known in mathematical modeling.

The next subsection gives some examples of hierarchical models.

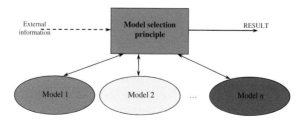

Fig. 5.17 The modern interpretation of a hybrid model. The multi-model approach

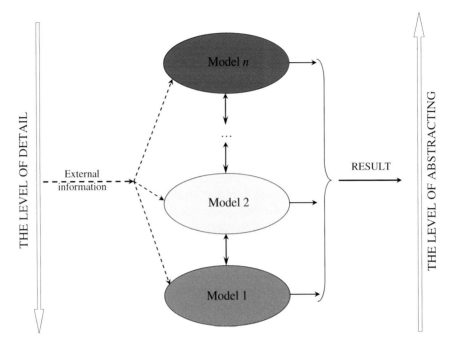

Fig. 5.18 A hierarchical (sequential) model

A Model of Warfare [153]

Suppose that opponents choose the "spatial" distribution of their forces (among springboards) one-time and simultaneously. In this case, we obtain *the colonel Blotto game* (CBG[15]), where the winner at each springboard results from solving the corresponding *Lanchester's equations*. In other words, it is possible to study an "hierarchical" model as follows. At the upper level, players allocate their forces among springboards within a certain variation of the game-theoretic model of the CBG. At the lower level, the result of a battle at each springboard is described by some modification of Lanchester's model. The complexity of such hierarchical models lies in that, in most cases, it is difficult to find the analytical solution to the CBG (see a survey in [107]).

In addition, Lanchester's models allow the hierarchical approach. At the lower level, *the Monte Carlo method* serves for simulating the interaction of separate military units. At the middle level, this interaction is described by *Markov models*.

---

[15]The classical CBG has the following statement. Two commanders (colonels Blotto and Lotto) distribute their forces among a finite number of springboards. The winner at each springboard is the player having more forces. Each commander strives for winning at as many springboards as possible.

Table 5.1 The model of warfare

Hierarchical level	Modeled phenomena/processes	Modeling tools
5	Spatial distribution of forces and means	The colonel Blotto game and its modifications
4	Temporal distribution of forces and means	Optimal control, repeated games, etc.
3	Size dynamics	Lanchester's equations and their modifications
2	"Local" interaction of units	Markov models
1	Interaction of separate military units	Simulation, the Monte Carlo method

And finally, the upper (aggregated, deterministic) level involves Lanchester's differential equations proper. By introducing control variables (temporal distributions of forces and means, reserves engagement, etc.), one can superstruct control problems "over" these models (in terms of controlled dynamic systems, *differential and/or repeated* games, etc.). Consequently, we obtain the following hierarchical model, illustrated by Table 5.1.

The Model of Distributed Penetration Through a Defense System (*The So-Called Diffuse Bomb Problem* [105])
An example of the hierarchical model of a MAS is the diffuse bomb problem stated below.

A group of autonomous moving agents must hit a target with given coordinates. At each time step, any agent can be detected and destroyed by a defense system (with a certain probability). Detection/annihilation probability depends on agent's coordinates and speed, as well as on the relative arrangement of all objects in the group. The problem is synthesizing algorithms of decentralized interaction among agents and their decision-making (the choice of direction and speed of their motion) to maximize the number of agents reaching the target. Agents appear "intelligent" in the following sense. Some agents (reconnaissance) can acquire on-line information on the parameters of the defense system. By observing the behavior of the reconnaissance agents, the rest ones perform "reflexion," assess the limits of dangerous areas and solve the posed problem. Strategic interaction of counteracting sides can be described in terms of game theory, see [106].

The following hierarchical model defined by Table 5.2 serves for appraising and choosing most efficient algorithms of behavior in [105].

The Hierarchical Structure of Agents in Multi-agent Systems (MAS)
In multi-agent systems (see Sects. 5.1 and 5.3), the hierarchy of models is *inter alia* generated by the functional structure of the agent. The latter may have several hierarchical levels, see Fig. 5.19 [153, 158]. The lowest (operational) level is responsible for implementation of actions (e.g., motion stabilization with respect to a preset path). Tactical level corresponds to actions' choice (including interaction with other agents). Strategic level is in charge of *decision-making, learning* and

Table 5.2 The diffuse bomb model

Hierarchical level	Modeled phenomena/processes	Modeling tools
6	Choosing the set of agents and their properties	Discrete optimization methods
5	Choosing the paths and speeds of agents	Optimal control
4	Agent's forecast of the behavior of other agents	Reflexive games. The reflexive partitions method
3	Detection probability minimization based on current information	Algorithms of course choice
2	Collisions avoidance, obstacles avoidance	Algorithms of local paths choice
1	Object's movement towards a target	Dynamic motion equations

adaptivity of behavior. And finally, the highest level (goal-setting) answers the principles of goal-setting and choice of the mechanisms of functioning for agents. The diffuse bomb problem realizes the general structure described by Fig. 5.19.

The structure presented by Fig. 5.19 seems rather universal. However, most realizations of multi-agent systems involve merely two lower levels and the framework of *dynamic systems theory*.

In *mission planning* problems, one can use different means of *artificial intelligence*, e.g., neural networks, evolutionary and logical methods, etc.

Also, let us mention *distributed optimization* (*agent-based computing*, see [32]) as a direction of modern optimization widespread in MAS. Its key idea consists in the following. An optimization problem of a multivariable function is decomposed into several subproblems solved by separate agents under limited information. For instance, each agent is "responsible" for a certain variable; at a current step, it chooses the value of this variable, being aware of the previous choice of some its "neighbors" and seeking to maximize its own local "goal function." Given an initial (global) goal function, is it possible to find the "goal functions" of agents and their interaction rules so that the autonomous behavior of agents implements a centralized optimum? (in *algorithmic/computational game theory* [4, 123], this optimum can correspond to a Nash equilibrium or a Pareto efficient state of agents' game).

Consider the strategic level of agent's architecture, which answers for adaptation, learning, reflexion and other aspects of strategic decision-making. *Game theory* and *theory of collective behavior* analyze interaction models for rational agents. In game theory, a common scheme consists in (1) describing the "model of a game," (2) choosing an equilibrium concept defining the stable outcome of the game and (3) stating a certain *control problem*–find the values of controlled "game parameters" implementing a required equilibrium (see Fig. 5.20, where "levels" correspond to the functions of science discussed in Sect. 1.1).

Taking into account *informational reflexion* leads to the necessity of constructing and analyzing awareness structures [158]. This enables defining an informational equilibrium, as well as posing and solving *informational control*

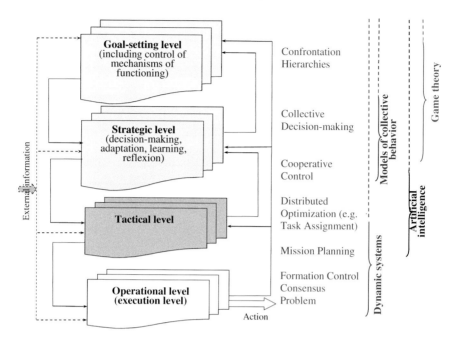

Fig. 5.19 The hierarchical structure of an agent in MAS

problems. Taking into account *strategic reflexion* generates a similar chain marked by heavy lines, i.e., posing and solving "*reflexive control*" problems [154].

The Model of Informational Confrontation in Social Networks

The object and means of control is a social network or another "networked" object [75, 89].

One can distinguish among several levels of description and analysis of social networks, see Table 5.3. At level 1 (the lowest one), a network is considered "in toto"; such description provides no details but is essential for rapid analysis of general properties enjoyed by the object. The aggregated description of a network employs *statistical methods*, *semantic analysis* techniques, etc.

Level 2 examines the structural properties of a network using the framework of *graph theory*.

The informational interaction of agents is analyzed at level 3; here we dispose of a wide range of applicable models (*Markov models*, *finite-state automata*, models of *innovations diffusion*, *infection models*, and others).

Level 4 involves *optimal control* or *discrete optimization* methods to formulate and solve control problems.

And finally, level 5 serves to describe the interaction of subjects affecting a social network (pursuing their individual goals). As a rule, this level utilizes *game theory* including reflexive games.

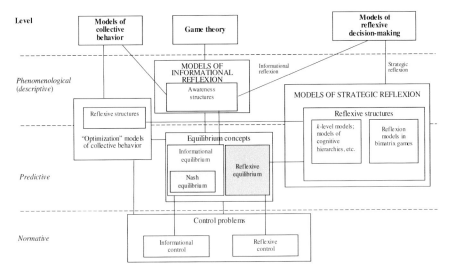

Fig. 5.20 Decision-making: informational and strategic reflexion

Consequently, we arrive at the following hierarchical model illustrated by Table 5.3.

The example of social media [75] highlights the problems of *social, economic and informational security in ICT*. Technological progress gradually increases its pace, and society appears unable to fully realize new opportunities and threats created by a certain technology. While discovering atomic power, scientists recognized possible problems in the case of its military application (e.g., recall the Einstein–Szilárd letter to the US President F.D. Roosevelt in 1939). Today, even experts have no totally clear understanding of the social impact of ICS. No doubt, ICT provide ample opportunities for decision-making, particularly, for expertise [73]. On the other hand, there arise new problems, too.

The results of functioning of computer-aided *decision support systems* (including the ones obtained within some formal models using modern ICT) are applied to make real important decisions. Hence, this aggravates security problems, i.e.,

Table 5.3 The model of informational confrontation in social networks

Hierarchical level	Modeled phenomena/processes	Modeling tools
5	Informational confrontation	Game theory, decision theory
4	Informational control	Optimal control, discrete optimization
3	Informational interaction of agents	Markov models, finite-state automata, models of innovations diffusion, infection models, etc.
2	Analysis of structural properties of a network	Graph theory
1	General analysis of a network	Statistical methods, semantic analysis techniques, etc.

making decisions and their consequences proof against the negative impacts of all the participating elements (both hardware components and active subjects).

Furthermore, society and government display growing interest in social media (online networks) as a source of specific information for predictive detection of aborning implicit tendencies to-be-controlled.

In other words, we inevitably face the problems of social, economic and informational security for an individual, society and a whole country: social, expert and other networks actually form an arena of informational contagion when control subjects struggle for the "minds" of other network members, whereas a social network itself represents an object and/or tool of informational impacts.

"Hierarchical Automation" in Organization-Technical Systems

Since the 1980s, *production systems* have followed a long path from flexible to holonic systems. In recent years, they attract the growing interest of researchers in connection with new market challenges: the efficiency of production specialization and decentralization, product and service differentiation, etc. There appear *networked productions* and *"cloud" productions*. Along with implementation of fundamentally new technologies of production (nanotechnologies, additive technologies, digital production, and so on), we observe gradual changes in its organization, i.e., the emphasis is shifted from operations automation to *control automation* at all life cycle stages.

Existing challenges such as:

- a huge number of product's customized configurations;
- integration of small- and large-scale production;
- lead-time reduction for an individual order;
- supply chains integration for stock optimization;

and others call for solutions guaranteeing:

- the universality of production systems and their separate components;
- the capability of rapid and flexible adjustment with respect to new tasks;
- autonomous decision-making in production owing to high-level control automation;
- survivability, replicability and scalability owing to network-centric control and multi-agent technologies;
- decision-making in production with proper consideration of economic factors, etc.

Modern production systems have a hierarchical structure, as indicated by Fig. 5.21. And the complexity of control problems treated induces their decomposition into decision-making levels. Each level in control problems solution corresponds to its own goals, *models* and *tools* (Fig. 5.21) at each stage of control (organizing, planning, implementing, controlling and analyzing). Hence, in organizational-technical production systems it is possible (and necessary) to apply hierarchical modeling.

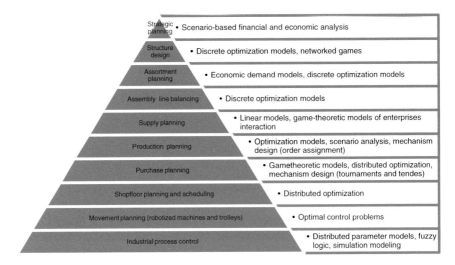

Fig. 5.21 Hierarchical models in production systems

This possibility is implemented, but on an irregular and unsystematized basis. Obviously, one can solve real problems of automation, analysis and decision support for production systems only within appropriate computer-aided informational systems. As an illustration, consider the classes of such systems in the ascending order of their "hierarchical level":

– lower-level control systems (PLC, MicroPC, …);
– supervising and scheduling systems (SCADA, DCS, …);
– production planning and management systems (MRP, CRP, …, MRP2, …);
– integrated systems (MES, …, ERP., …);
– systems responsible for interaction with an external environment or development (SCM, CRM, PMS, …);
– upper-level analytic systems (OLAP, BSC, DSS, …).

These classes of systems use mathematical models, but very sparsely; as a rule, the higher is the level of hierarchy,[16] the lesser is their usage. For instance, lower-level controllers employ in full automatic control theory; project management systems (PMS) incorporate classical algorithms for critical path search, Monte Carlo methods for project duration estimation, and heuristics for resources balancing; ERP systems and logistics systems (SCM) involve elementary results from stock management theory, and so on.

Nevertheless, full-fledged implementation of the so-called "*hard*" *models* and "quantitative science" (operations research, discrete optimization, data analysis and

---

[16]This statement is true for separate informational systems and for integrated informational systems of product life cycle management (PLM) including computer-aided design systems, which realize the complex of the listed functions.

other branches of modern applied mathematics) in informational systems still waits in the wings.

Several global problems exist here. On the one hand, mathematical models require very accurate and actual information often associated with inadmissibly high organizational and other costs. On the other hand, in many cases "*soft*" *models* (putting things in order in production processes, implementation of typical solutions and standards in the form of qualitative best practices, etc.) yield an effect exceeding manyfold the outcomes of quantitative models, yet consume reasonable efforts. Therefore, it seems that quantitative models should be applied at the second stage, "extracting" the remainder of potential efficiency increase.

Concluding this section dedicated to heterogeneous models and hierarchical modeling, we underline a series of their common classes of problems. Modern controlled objects are complicated so that sometimes a researcher would hardly separate out purely hierarchical or purely networked components. In such cases, it is necessary to consider *networks of hierarchies* and *hierarchies of networks*.

First, at each level models have their own intricacies induced by a corresponding mathematical apparatus. Moreover, there arise "conceptual coupling" dilemmas and the *common language* problem among the representatives of different application domains.

Second, a complex of "*joined*" *models* inherits all negative properties of each component. Just imagine that, at least, one model in a "chain" admits no analytic treatment; then the whole chain is doomed to simulation modeling. The speed of computations in a chain is determined by the slowest component, and so on.

And third, it is necessary to assess the comparative efficiency of the solutions of aggregated problems, as well as to elaborate and disseminate typical solutions of corresponding control problems in order to transfer them to the engineering ground.

5.5 Strategic Behavior

Control theory has followed a long path of development from automatic regulation systems to *intelligent control* systems, as illustrated by Figs. 5.11 and 5.22.

Intelligent control can be defined in different ways, namely, as control including goal-setting [211]; as control based on artificial intelligence methods (e.g., artificial neural networks, evolutionary (genetic) algorithms, logical inference or logical and dynamic models, knowledge representation and knowledge management, etc.[17]); as

---

[17]Each of these classes possesses certain advantages and shortcomings, especially, in the sense of real-time requirements. Today, the choice of concrete tools is defined by the skill of a researcher or engineer, as well as by accumulated experience and traditions of corresponding scientific schools. Global challenges concern maximum suppression of existing shortcomings of separate tools and design of general methods for their integration subject to posed problems.

Fig. 5.22 From automatic regulation to intelligent control

control imitating human behavior, and so on. Not all "definitions" seem appropriate.

Unfortunately, the term "intelligent" has become a fashionable attachment to control system (behavior, etc.) description, and the absence of such characteristic is interpreted as being out-of-date. This "devalues" the whole essence of intelligence.

In the previous sections, we have identified several properties of interdisciplinary-nature systems comprising human beings (or artificial systems "imitating" human beings) such as independent goal-setting, purposeful behavior, reflexion, bounded rationality, cooperative and/or competitive interaction. All these properties can be covered by the category of *strategic behavior*. From the historical perspective, systematic consideration of the human factor (including strategic behavior of a controlled object) in mathematical control problems was pioneered in *theory of active systems* in the late 1960s. This theory was founded by V. Burkov, see the first publications [36, 39], the survey [37] and the modern textbooks and monographs [38, 131].

Further exposition focuses on some actual aspects of strategic behavior. However, we will not discuss many "internal" problems of associated scientific directions such as game theory, mechanism design, and others.

Intelligent multi-agent systems. One modern tendency in *theory of multi-agent systems*, *game theory* and *artificial intelligence* lies in that researchers strive for their integration. Yet, game theory and artificial intelligence aim at higher levels of agent's architecture, see Fig. 5.19.

Within the so-called *algorithmic game theory* [4], one would observe "transition downwards" (see Fig. 5.23), i.e., from the uniform description of a game to its decentralization and analysis of the feasibility of implementing autonomously the mechanisms of equilibrium behavior and realization. On the other hand, theory of MAS moves "upwards" (see Fig. 5.23) in a parallel noncoincident way due to the local character of scientific communities. Theory of MAS aspires after better consideration of *strategic behavior* and design of typical *test problems and scenarios*. The latter are necessary, since in most cases tactical level employs certain heuristic algorithms to-be-compared in terms of complexity, efficiency and other criteria (the number of heuristic algorithms demonstrates rapid growth owing to intensive research of multi-agent systems).

The concept of *bounded rationality* gradually becomes widespread in analysis (perhaps, this tendency will be even stronger in future): in the absence of time, possibility or vital necessity, investigators search for admissible pseudo-optimal control actions instead of optimal ones (in many situations, on the basis of heuristic methods).

Fig. 5.23 MAS and strategic
behavior: state-of-the-art

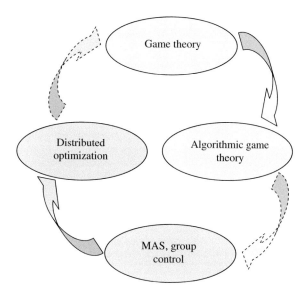

Furthermore, consideration of the human factor calls for employing *mechanism design* [131] and behavioral theories (*experimental economics, experimental game theory*, see a survey and references in [155]). The "normative" picture of interaction between MAS and strategic behavior sciences has the form demonstrated by Fig. 5.24.

In addition, we emphasize that aspiration for maximum intellectualization is bounded by "costs" (computational, cognitive, tactical and technical, economic and other costs), see Fig. 5.25. In other words, in MAS agents must have a *rational "intellectualization"* level being adequate to a posed problem in terms of "costs." On the other hand, aspiration for maximum intellectualization as maximization of the guaranteed efficiency of MAS functioning over the set of feasible situations at goal-setting level corresponds to decentralization rejection, i.e., transition to a centralized system.

Reflexion. *Game theory* studies interaction of superintelligent agents with same cognitive capabilities as their researcher [141], whereas *theory of collective behavior* proceeds from agents' rationality (or bounded rationality). A possible bridge[18] between them for transition from rational to superintelligent agents consists in increasing agents' "intellectualization" by endowing them with *reflexive capabilities*, see Fig. 5.26 and surveys in [154, 158].

Informational reflexion is the process and result of agent's thinking about (a) the values of uncertain parameters and (b) what its opponents (other agents) know about these values. Here the "game" component actually disappears—an agent makes no decisions.

---

[18]Alternatives are, e.g., consideration of evolutionary games [220] or learning effects in games [141].

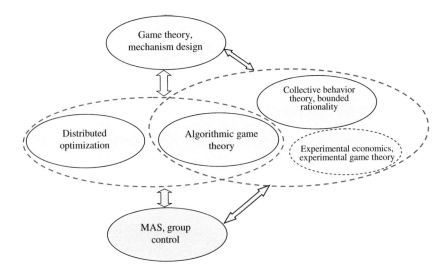

Fig. 5.24 MAS and strategic behavior sciences: the normative picture of interaction

Fig. 5.25 The price of "intellectualization"

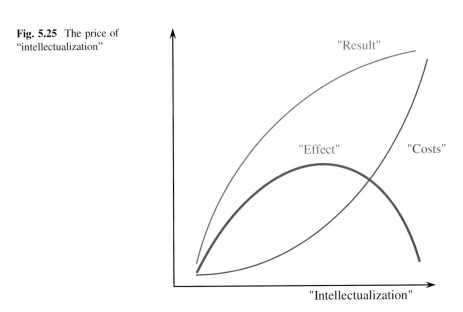

Strategic reflexion is the process and result of agent's thinking about which decision-making principles its opponents (other agents) employ under the awareness assigned by it via informational reflexion, see Fig. 5.20.

A key role belongs to the notions of *informational/reflexive structures* describing the nontrivial mutual awareness of agents (or their self-awareness, see ethical

Fig. 5.26 Reflexion and growing "intellectualization"

choice models in [115]) and *phantom agents* existing in the minds of other real and phantom agents and possessing certain awareness.

The concept of phantom agents yields rigorous statement of *reflexive games* as games of real and phantom agents (the term suggested in 1965 by V. Lefevbre [116]). Moreover, this concept allows defining *informational equilibria* as a generalization of Nash equilibria for reflexive games: each (real or phantom) agent evaluates its subjective equilibrium (an equilibrium in a game this agent thinks it actually plays) based on an existing hierarchy of believes about the objective and reflexive realities [158].

Reflexive games research yields the following. First, it provides a uniform methodology and mathematical framework to describe and analyze various situations of collective decision-making by agents possessing different awareness, to study the impact of reflexion ranks on agents' payoffs, to obtain conditions of existence and implementability of informational equilibria, etc. Second, such research makes it possible to establish the existence conditions and properties of an informational equilibrium, as well as to pose constructively and correctly the *problem of informational control*. In this problem, a Principal has to find an awareness structure such that the informational equilibrium implemented in it appears most beneficial to it. An interested reader can find a necessary theoretical background and numerous applications of reflexive games and informational control in the book [158].

The achievements and illusions of "emergent intelligence." This section ends with a brief consideration of a phenomenon related to "intelligent" control and behavior of artificial (e.g., multi-agent) systems.

In the two recent decades, much attention of researchers in cybernetics and artificial intelligence has been paid to *emergent intelligence*. A system composed of very many relatively simple homogeneous elements (e.g., agents in MAS[19]) locally interacting with each other and an external environment demonstrates a complex[20] "intelligent" behavior in comparison with the simplicity of its elements. Investigations in this field are also motivated by existing analogs in nature (*Swarm Intelligence*, i.e., heuristic algorithms of distributed optimization in ant colonies and beehives, flocks of birds, fish shoals, etc.).

---

[19]This class also includes the problematique of artificial neural and immune networks, probabilistic automata, genetic algorithms, and so on.

[20]Some authors insist on the birth of a new science called complexity science.

Such systems enjoy a series of obvious advantages: the cheapness and simplicity of a separate element, local fault-tolerance, scalability, reconfigurability, asynchrony, parallel processing of local information (*ergo*, high-level performance of *real-time* operation). They have numerous applications: social systems (crowd wisdom, e-expertise, social networks, etc.), economic systems (financial and other markets, national and regional economics, etc.), telecommunication networks, models of production and transport logistics systems, robotics, knowledge extraction (particularly, from Internet), Internet of Things and others [53, 73, 75, 151, 183, 195].

The appearance of qualitatively new properties in a whole system (against the individual properties of its elements), i.e., transition from simple local and decentralized interaction of elements to a nontrivial and complex global behavior, allows treating the latter as *adaptive* and *self-organizing*. Indeed, **nonlinearity, evolution, adaptivity and self-organization are the characteristic features of real modern complex systems** (e.g., see examples and their discussion in [183]).

In addition to many achievements and good prospects, emergent intelligence sometimes creates several illusions. Actually, emergent intelligence concerns artificial systems, but adaptation and self-organization (despite all their pluses) are embedded at the stage of system design. Notwithstanding *the law of emergence* (the whole is greater than the sum of its parts, see above), the behavior of artificial systems gets predetermined by the behavior/interaction of its elements.

Similar delusions occurred in the history of science (e.g., at the early development stages of cybernetics and artificial intelligence[21]). They induced much disappointment and put the brakes on the evolution of these scientific directions.

Furthermore, recall that MAS realize heuristics and it is necessary to assess the guaranteed efficiency of their solutions, see above.

Generally speaking, there exist three large sources of "new" properties of a system:

– *additive interaction*[22] of its elements;
– for an observer/researcher having *limited information* and *cognitive capabilities*, *the multiplicity of elements and their mutual relations* (perhaps, nonlinear, asynchronous, with delayed information exchange, etc.) makes it impossible to conduct a mental experiment for reproducing agents' behavior in detail; and computer simulation yields "surprising"[23] results (an unexpected system behavior);

---

[21]A cybernetical system always has the behavior defined by its embedded algorithms ("stochastic," "nondeterministic," and others), despite the seeming generation of new knowledge or demonstration of qualitatively new ("unexpected") behavior. This is especially the case under interaction of very many elements (a simple-structure system shows a complex behavior).

[22]For instance, a microrobot cannot move a heavy load, in contrast to many microrobots applying their joint efforts.

[23]The complete model of a system is so complicated that the appearance of new properties represents a "miracle" for an external observer (at the same time, scientists intensively exploit it and start believing that an artificial system can demonstrate an "independent" behavior).

– *artificial randomization* (embedded into behavioral algorithms describing agents' interaction with each other and/or an external environment) is necessary for variety creation (in the final analysis, for self-organization).[24]

5.6   Big Data and Big Control

In information technology, *big data* represents a direction of theoretical and practical investigations on the development and application of handling methods and means for the big volumes of unstructured data. Perhaps, the term was first mentioned in the special issue of *Nature* [144].

Big data handling comprises their[25]:

– acquisition;
– transmission;
– storage (including recording and extraction);
– processing (transformation, modeling, computations and analysis);
– usage (including visualization) in practical, scientific, educational and other types of human activity.

In the narrow interpretation, the term "big data" sometimes covers only the technologies of their acquisition, transmission and storage. In this case, big data processing (including construction and analysis of corresponding models) is called *big analytics* (including big computations), whereas visualization of the corresponding results (depending on user's cognitive capabilities) is called *big visualization* (see Fig. 5.27).

The universal cycle of big (generally, any) data handling is illustrated by Fig. 5.28. Here the key role belongs to an *object* and a *subject* (a "customer"); the latter requires knowledge on the state and dynamics of the former. However, sometimes there exists a chasm between *data* acquired on an object and *knowledge* necessary for a subject. Primary data must be preprocessed, i.e., transformed into more or less structured *information*. Subsequently, necessary knowledge is extracted from this information depending on a specific task solved by a subject.

---

[24]An uncertainty is always induced by some other uncertainty potentially comprising lack of knowledge (insufficient information) and/or the action of random factors (an uncertainty never arises from an abstract "complexity" and similar conceptual factors). Facing an "uncertainty," one should analyze cause-and-effect relations and seek for its source ("initial uncertainty"). Of course, different complexity factors merely get the things into muddle.

[25]In some classifications, big data handling is associated with 4D (data discovery, discrimination, distillation and delivery/dissemination).

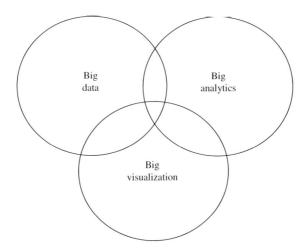

Fig. 5.27 "The big triad": data, analytics, visualization [*We will not discuss another fashionable triad* (*big data, high-performance computations, cloud technologies*)]

Particularly, a subject may adopt this knowledge for object *control*, viz., exerting purposeful impacts on an object to ensure its required behavior. Control can be automatic in a special case (an inanimate subject). Perhaps, the term "***big control***"[26] will become common soon for indicating control based on big data, big analytics and, possibly, big visualization,[27] see [151].

The overwhelming majority of big data investigations create the technologies of big data acquisition, transmission, storage and preprocessing, whereas big analytics and visualization receive by far less consideration. However, the emphasis is gradually shifted towards efficient algorithms of big data handling.

Sources and "customers" of big data:

– science (astronomy and astrophysics, meteorology, nuclear physics, high-energy physics, geoinformation systems and navigation systems, distant Earth probing, geology and geophysics, aerodynamics and hydrodynamics, genetics, biochemistry and biology, etc.);

– Internet (in the wide sense, including *Internet of Things*) and other telecommunication systems;

---

[26]As we have mentioned above, in the recent 15 years experts in control theory have tended to consider the problems of control, computations and communication jointly (the so-called C^3 problem (Control, Computation, Communication)). According to this viewpoint, control actions are synthesized in real time taking into account the existing delays in communication channels and information processing time (including computations). There is another generally accepted term (large-scale systems control), but big data can be generated by "small" systems.

[27]An alternative interpretation of "big control" concerns control of big data handling processes. Actually, this represents an independent and nontrivial problem.

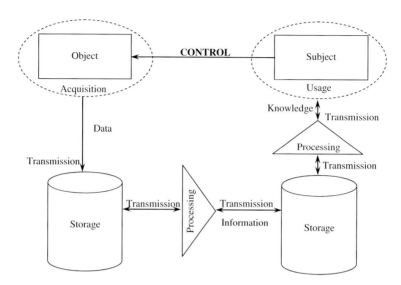

Fig. 5.28 The universal cycle of big data handling

- business, commerce and finances, as well as marketing and advertising (including trading, targeting and adviser systems, CRM-systems, RFID–radiofrequency identifiers used in sales, transportation, logistics and so on);
- monitoring (geo-, bio-, eco-; space, air, etc.);
- security (military systems, antiterrorist activity, etc.);
- power engineering (including nuclear power engineering), SmartGrid;
- medicine;
- governmental services and public administration;
- production and transport (objects, units and assemblies, control systems, etc.).

Numerous applications[28] of big data in these fields can be found in popular science literature (or even "glossy" journals) available at public Internet sources. We will not describe these applications here to avoid embarrassing "zettabytes" and "yottabytes."

In almost all fields cited, the modern level of automation is such that big data have automatic generation. Therefore, the following question gains growing importance. What is the volume of "lost" data flows (due to insufficient capabilities or time for their storage or processing)? This question seems correct for an engineer in ICT, but not for a scientist or a user of big data processing results. Rather, the former and the latter would ask "What are essential losses in this case?" and "What are the changes if we successfully acquired and processed all data?", respectively.

---

[28]The principal idea of using big data is revealing "implicit regularities," i.e., answering nontrivial questions: epidemic prediction based on information from social networks and sales in drugstores; medical and technical diagnostics; retention of clients by analyzing sellers' behavior in stores (the spatial movements of RFID-tags of products); and others.

Traditionally, big data are unstructured data whose volume exceeds the available handling capabilities in required time. However, this definition appears somewhat "cunning": data considered big today cease to be such tomorrow owing to the progress of data handling methods and means. Data that looked big several hundreds or even thousands of years ago (in the absence of automatic treatment) are easily processed today by home computers. The competition between the (hypothetic) computational demands of mankind and corresponding technical capabilities had been known very long ago. Of course, the capabilities have been always chasing the needs. And the gap between them represents a monumental stimulus for science development. Researchers have to suggest simpler (yet, adequate) models, design more efficient algorithms, etc.

Sometimes, the definition of big data includes the so-called 5V properties (Volume, Velocity, Variety, Veracity, Validity). Alternatively, the difference between the big volume of conventional data and big data proper is that the latter form the big flow of unstructured[29] data (in the sense of volume and velocity as the volume per unit time).

In the wide comprehension, the unstructuredness of big data (text, video, audio, communications structures, etc.) is actually their characteristic feature and a challenge for applied mathematics, linguistics, cognitive sciences and artificial intelligence. Creation of real-time processing technologies,[30] including the feasibility of implicit information revelation, for large flows of text, audio, video and other information forms the mainstream of applications of the above sciences[31] to ICT.

Therefore, we observe a direct (and explicit) query from technologies to science. The second explicit query concerns adaptation of traditional statistical analysis, optimization and other methods to big data analysis. Furthermore, it is necessary to develop new methods with due consideration of big data specifics. A modern fashionable trend is boosting analytics tools (generally, business analytics) for big data. But their list almost coincides with the classical kit of statistical tools (or is even *narrower*, since some methods are inapplicable to big data). This is also the case for:

– *machine learning methods* (support vector machine, random forests, artificial neural networks, Bayesian networks including separation of informational attributes and dimension reduction of attribute spaces in model relearning) and *artificial intelligence methods*;
– *high-dimensional optimization problems* (in addition to traditional parallel computing, intensive research focuses on distributed optimization);

---

[29]Data unstructuredness can be the result of their omissions and/or different scales of studied phenomena and processes (in space and time, see the so-called multi-scale systems).

[30]In the first place, these technologies must perform data aggregation (e.g., detecting changes in technological data or storing aggregated indices). Really, one does not need all data (especially, "homogeneous" data).

[31]Mathematics rather easily operates structured data; and so, data structuring makes an important problem.

– *discrete optimization methods* (here an "alternative" lies in application of multi-agent program systems–see the above discussion of distributed optimization problems).

The common feature in the stated queries of technologies to science is the insufficiency of adaptation or small modification of well-known tried-and-true methods. We have to be aware of the following. Generally, automatic modeling (by traditional tools[32]) based on raw data represents just a fashionable delusion.[33] We expect to suggest algorithms and apply them to bulky volumes of unstructured (often irrelevant) information, thereby improving the efficiency of decision-making (recall the "emergent intelligence illusion"). There exist no miracles in science: generally, new conclusions require new models and new paradigms (e.g., see the books on science methodology [112, 149]).

The complexity of the surrounding world grows at a smaller rate than the capabilities of data detection ("measurement") and storage. Perhaps, these capabilities have exceeded the ability of mankind to realize the feasibility and reasonability of their usage. In other words, we "choke" with data, trying to find what to do with them.

However, there exists an alternative viewpoint of this situation as follows. Obtaining *big* data (having an arbitrary large volume) is possible and easy enough (obvious examples arise in *combinatorial optimization, nonlinear dynamics* or *thermodynamics*, see below). But we have to understand how to manage big data (and ask the Nature correct questions). Furthermore, it is possible to construct an arbitrary complex model using big data and then try to reach a higher accuracy within the model. But the associated dilemma is whether we obtain new results or not (in addition to very many new problems[34]). Long ago mathematicians and physics knew that increasing the dimensionality and complexity of a model (aspiration for considering more factors and relations among them) does not necessarily improve the quality of modeling results; sometimes, it even carries to the point of absurdity.[35]

Based on analysis of several examples, the paper [151] distinguished between *natural* and *artificial* big data depending on their source. In the former case, data are generated by some independent object and we ("investigators") decide what should be "measured." In the latter case, the source of big data is a model; complexity (data flow) is partially controlled and defined during simulation.

---

[32]An additional encumbrance is the accumulated experience of a researcher/developer and the traditions of his scientific school. Successful solution of a certain problem leads to the conviction that same methods (only!) are applicable to the rest open problems.

[33]In some cases, additional information can be obtained by increasing the volume of data (under correct processing).

[34]We recognize the importance of model's adequacy and stability of modeling results, but omit these problems.

[35]Not to mention situations, when existing scientific paradigms make it impossible in principle to model system behavior on a large time horizon (e.g., accurate weather forecasting).

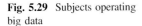 **Fig. 5.29** Subjects operating big data

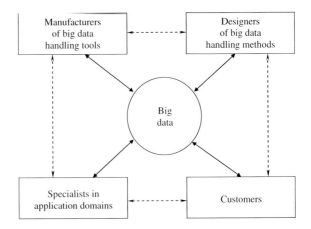

"Recipes." There exist four large groups of subjects (see Fig. 5.29) operating (explicitly or implicitly) big data in their professional (scientific and/or practical) activity:

- manufacturers of big data handling tools (software/hardware developers, suppliers, consultants, integrators, etc.);
- designers of big data handling methods (experts in applied mathematics and computer science);
- specialists in application domains (scientists focused on real objects or their models) that represent big data sources;
- customers utilizing or planning to utilize the results of big data analysis in their activity.

Representatives of the mentioned groups interact with each other (see the dashed lines in Fig. 5.30). The normative ("ideal") division of "responsibility areas" is illustrated by Fig. 5.30; here the thickness of arrows corresponds to the level of involvement.

Proceeding from *sensus communis* and not claiming to be constructive, we formulate the following general "recipes" for the listed groups of subjects.

For manufacturers of big data handling tools: with the course of time, it will be difficult to sell big data solutions (including analytical ones) without suggesting new adequate mathematical methods and stipulating for the feasibility of a close cooperation between customers, the developers of appropriate methods and specialists in application domains.

For mathematicians (the author's "brothers-in-arms"): a topical query concerns adapting well-known methods and developing new processing methods (in the first place, with nonlinear complexity!) for the large flows of unstructured data

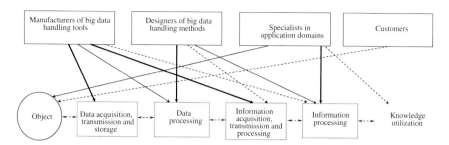

Fig. 5.30 The division of "responsibility areas"

representing a good testing area for new models, methods and algorithms (to the extent possible, at the expense of manufacturers and/or customers).

For specialists in application domains: big data technologies lead to new capabilities for acquiring and storing the bulky arrays of "experimental" information, conducting the so-called computing experiments; the associated methods of applied mathematics enable systems generation and rapid verification of hypotheses (revelation of implicit regularities).

For customers: the expensive technologies of big data acquisition and storage would hardly be economically sound without involving specialists in appropriate methods and subject areas (only if it is absolutely clear which questions a customer would like to answer using big data[36]).

As a positive trend in big data handling, note aspiration for seeking adequate *macrodescriptions of big systems.* For instance, consider research works on social systems modeling, i.e., social networks, mob and so on, which involve microdescriptions (at the level of separate agents) [18] and macrodescriptions (in terms of distribution functions of essential parameters) [34], as well as establish a correspondence between them [33]. Such approach is also developed within the framework of *sociophysics* and *ecophysics*, where *statistical physics* tools are applied to model *complex networks* and big socioeconomic systems.

Some threats. In addition to the emphasized necessity of searching for adequate simple models and the alerting trend of anticipatory technologies development, we expect the future relevance of the following problems (the list below is unstructured and incomplete).

- *The informational security of big data.* This requires adaptation of well-known methods and tools, as well as development of fundamentally new ones. Really, alongside with the growing topicality of *cybersecurity problems* (in the wide sense, the informational security of control systems) and the problem of security "against information" (especially, in social networks), one should consider the specifics of big data proper.

---

[36]Though, it is possible to store data de bene esse (e.g., to verify a certain hypothesis in future based on them).

- *The energy efficiency of big data.* Even today, data processing centers represent a considerable class of power consumers. The bigger are data to-be-processed, the higher is energy needed.
- *The principle of complementarity* was established in physics long ago; it declares that measurements modify the state of a system. However, does it apply to social systems whose elements (people) are active, i.e., possess their own interests and preferences, choose their actions independently, etc. [36, 131, 157]?

A demonstration of this principle lies in the so-called *information manipulation* (*strategic behavior*). According to theory of choice [36, 38, 39, 131], an active subject reports information by forecasting the results of its usage; generally speaking, an active subject does not adhere to truth-telling.

Another example concerns the so-called *active forecasting*: a system changes its behavior based on new knowledge about itself [158].

Are these and similar problems eliminated or aggravated in the case of big data?

- Recall *the principle of uncertainty* in the following (epistemological) statement [149]: the current level of science development is characterized by certain mutual constraints imposed on results "validity" and results applicability, see Fig. 5.11. In the context of big data, this principle means the existence of a rational balance between the level of detail in the description of a studied system and the validity of results and conclusions to-be-made on the basis of this description.
- A traditional assumption in design and operation of information systems (corporate systems, decision support systems of governmental services, inter-agency circulation of documents, etc.) is that all information in such systems must be complete, unified and publicly available (under existing access rights). But it is possible to show the "distorting-mirror" reality to each person, i.e., to create an individual informational picture,[37] thereby performing *informational control* [157, 158]. Should we strive for or struggle against these effects in the field of big data?

Summarizing the above consideration of trends and forecasts in control theory, we declare that a similar (or even more systematic, regular and in-depth) analysis is vital for other sciences, viz., cybernetics, systems analysis, optimization, artificial intelligence, etc. This would give an impetus for the evolution of Cybernetics via the appearance of new generalizations in the form of corresponding laws, regularities, principles and so on.

Educational support. Concluding this chapter, we discuss a partial (yet, important) aspect of the modern state of control theory, namely, its educational support.

---

[37]At the very least, a fragment of the "objective" picture (hushing up the whole truth); at the most, an arbitrary inconsistent system of beliefs about the reality.

Let us appeal to readers considering themselves as experts in control theory with the following dilettantish request[38]: "Please, recommend a *textbook* on modern control theory (a one-year course not restricted to *automatic control theory* (ACT) or even to linear systems, robust control or another branch of control theory) so that an uninitiated student specializing in mathematics or engineering would form a complete and, conversely, superficial notion of modern control theory."

Unfortunately, the request leads to deplorable results. On the one part, there are good reference books [200], textbooks and handbooks on ACT, both classical (e.g., see [167, 215] and a survey in [83]) and modern ones (e.g., see [3, 16, 35, 51, 161]). On the other part, excellent textbooks and monographs focus on separate branches of ACT: robust control [171], nonlinear systems control [96] and others.

And so, modern textbooks and reference books provide a good coverage of classical ACT, but almost ignore general statements of control problems and decision-making problems (being confined to dynamic systems as a "universal" descriptive framework for any controlled objects) and pay little attention to intelligent control, networked control, the "sectoral" specifics of controlled objects and so on. To our regret and despite the efforts of N. Wiener and its followers on creation of a universal control science, none of textbooks on ACT deeply treats the generality of laws and processes of control in the animal, machine and society.

Imagine that this request ("Please, recommend a *textbook* on modern control theory...") is addressed to a potential reader without well-developed skills in higher mathematics (e.g., a schoolchild). How can we make the results of modern control theory clear to such readers? Here the situation seems even worse. Of course, (a) the amount of scientific knowledge accumulated in control theory is huge, (b) the study of this knowledge requires special training, (c) a dilettante would never perceive it, (d) the described function is performed by handbooks and reference books, ... But a counterargument is that today many sciences (physics, chemistry, biology) can be presented at the levels of a school textbook, a university textbook or a scientific monograph. For instance, such "encyclopaedic" textbooks exist for other "capacious" sciences, namely, informatics, artificial intelligence, game theory, operations research, etc. Why are there no school textbooks on control theory[39] and only a few broad university textbooks? Creation of easy-to-understand (yet, rigorous and complete) textbooks on control theory is an urgent challenge for experts in the field!

---

[38]Another "educational" question ensuing from the generality of control laws and principles can be stated as follows: "Is it better to organize a department for control problems in each "sectoral" university or a university dedicated to control problems with "sectoral" departments?". The book will touch this question.

[39]Speaking about "control theory," we mean exactly mathematical control theory (and not a corresponding branch of management science discussed in numerous bélles-léttres textbooks available today at stores).

Conclusion
Cybernetics 2.0

Therefore, we have briefly considered the history of cybernetics and its state-of-the-art, as well as the development trends and prospects of several components of cybernetics (mainly, control theory). What are the prospects of cybernetics? To answer this question, let us address the primary source—the initial definition of cybernetics as the science of CONTROL and COMMUNICATION.

Its interrelation with control seems more or less clear. At the first glance, this is also the case for communication: by the joint effort of scientists (including N. Wiener), the mathematical theory of communication and information appeared in the 1940s (quantitative models of information and communication channels capacity, coding theory, etc.).

But take a broader view of communication.[1] Both in the paper [181] and in the original book [221], N. Wiener explicitly or implicitly mentioned interrelation or intercommunication or interaction—*reasonability* and *causality* (*cause-effect relations*). Really, in *feedback control systems*, control-effect is defined by its cause, i.e., the state of a controlled system (plant); conversely, control supplied to the input of a plant is induced by its cause, i.e., the state of a controller, and so on. No doubt, the channels and methods of communication are important but secondary whenever the matter concerns universal regularities for animals, machines and society.

A much broader view of communication implies interpreting communication as INTERCOMMUNICATION, e.g., between elements of a plant, between a controller and a plant, etc. including different types of impacts and interactions (material, informational and other ones). "Intercommunication" is a more general category than "communication."

---

[1]Academician A. Kolmogorov was against such interpretation. In 1959 he wrote: "Cybernetics studies any-nature systems being capable to perceive, store and process information, as well as to use it for control and regulation. Cybernetics intensively employs mathematical methods and aims at obtaining concrete special results, both in order to analyze such systems (restore their structure based on experience of their operation) and to design them (calculate schemes of systems implementing given actions). Owing to this concrete character, cybernetics is in no way reduced to the philosophical discussion of reasonability in machines and the philosophical analysis of a circle of phenomena explored by it." We venture to disagree with this opinion of a great Soviet mathematician.

© Springer International Publishing Switzerland 2016
D.A. Novikov, *Cybernetics*, Studies in Systems, Decision and Control 47,
DOI 10.1007/978-3-319-27397-6

In the general systems context, intercommunication corresponds to the category of ORGANIZATION (see its definition and discussion below). Therefore, a simple correction (replacing "communication" with "organization" in Wiener's definition of cybernetics) yields a more general and modern definition of cybernetics: "the science of systems organization and their control." We call it *cybernetics 2.0.*

Making such substitution, we get distanced from informatics. Consider the soundness and consequences of this distancing.

Cybernetics and informatics. Nowadays, cybernetics and informatics form independent interdisciplinary fundamental sciences [101]. According to a figurative expression of Sokolov and Yusupov [191], informatics and cybernetics are "Siamese twins." Yet, in nature Siamese twins represent pathology.[2]

Cybernetics and informatics have a strong intersection (including the level of common scientific base—statistical information theory[3]). Their accents much differ. The fundamental ideas of cybernetics are Wiener's "control and communication in the animal and the machine," whereas the fundamental ideas of informatics are formalization (theory) and computerization (practice). Accordingly, in the mathematical sense cybernetics bases on control theory and information theory, whereas informatics proceeds from theory of algorithms and formal systems.[4]

The subject of modern informatics (or even the "umbrella brands" of *informational sciences*) covering information science, computer science and computational science [102] are informational processes.

Indeed, on the one hand, information processing arises everywhere (!), not only in control and/or organizing. On the other hand, informational processes and corresponding information and communication technology are integrated into control processes[5] so that their discrimination seems almost impossible. A close cooperation of informatics and cybernetics at partial operational level will be continued and even extended in future.

Organization. Organization theory. Organizational culture. According to the definition provided by Merriam-Webster dictionary, *an organization* is:

---

[2]For instance, the definition of informatics as the "union" of general laws of informatics and control would induce a megascience without concrete content, subsisting at conceptual level exclusively.

[3]Note that mathematical (statistical) theory of communication and information operates quantitative assessments of information. Unfortunately, no essential advancements have been made in the field of substantial (semantic) value of information. This problem is still a global challenge of informatics.

[4]This distinction partly elucidates why some sciences often related to informatics or computer sciences have not been reflected in the book: theory of formal languages and grammars, "true" artificial intelligence (knowledge engineering, reasoning formalization, behavior planning, etc. instead of artificial neural networks as a modern empirical engineering science), automata theory, computational complexity theory, and so on.

[5]N. Wiener believed that control processes are, in the first place, informational processes: information acquisition, processing and transmission (see the above discussion of joint solution of problems appearing in control, computations and communication).

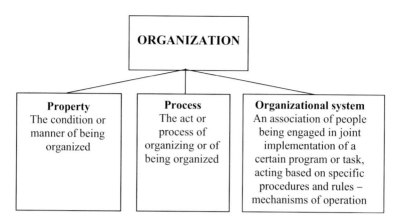

Fig. A.1 Definition of organization

1. The condition or manner of being organized;
2. The act or process of organizing or of being organized;
3. An administrative and functional structure (as a business or a political party); also, the personnel of such a structure—see Fig. A.1.

The present book uses the notion "organization" mostly in its second and first meanings, i.e., as a process and a result of this process. The third meaning (an organizational system) as a class of controlled objects appears in theory of control in *organizational systems* [131, 157].

At descriptive (phenomenological) and explanatory levels, "system organization" reflects HOW and WHY EXACTLY SO, respectively, a system is organized (organization as a *property*). At normative level, "system organization" reflects how it MUST be organized (requirements to the *property* of organization) and how it SHOULD be organized (requirements to the *process* of organization).

A scientific branch responsible for the posed questions (*Organization*[6]*theory*, or **O³** (**organization as a property, process and system**, by analogy to C³ as discussed above) has almost not been developed to-date. Yet, this branch obviously has a close connection and partial intersection with general systems theory and systems analysis (mostly focused on descriptive level problems and a little bit dealing with normative level ones), as well as with methodology (as the general science of activity organization [148]). **Creating a full-fledged Organization theory is a topical problem of cybernetics!**

---

[6]Note that there also exists "theory of organizations" ("organizational theory")—a branch of management science, both in its subject (organizational systems) and methods used. Unfortunately, numerous textbooks (and just a few monographs!) give only descriptive generalizations on the property and process of organization in their Introductions, with most attention then switched to organizational systems, viz., management of organizations (for instance, see the classical textbooks [47, 134]).

Table A.1 Types of organizational culture: a characterization [148, 152]

The types of organizational culture	The methods of normalization and translation of activity	The forms of social structure implementing the corresponding method
Traditional	Myths and rituals	Communities based on the kinship principle
Corporate-handicraft	Samples and recipe for their recreation	Corporations with a formal hierarchical structure (masters, apprentices, and journeymen)
Professional (scientific)	Theoretical knowledge in the form of text	Professional organizations based on the principle of ontological relations (relations of objective reality)
Project-technological	Projects, programs and technologies	Technological society being structured by the communicative principle and professional relations
Knowledge-based	(Individual) and collective knowledge about activity organization	Networked society of knowledge

Speaking about the notion of organization, one should not ignore the phenomenon of organizational culture. Different historical periods of civilization evolvement are remarkable for different types of activity organization now called *organizational culture*, see Table A.1.

Presently, the *knowledge-based type of organizational culture* gradually manifests itself. Here exactly (individual and collective) knowledge about activity organization (!) is the product and way of activity normalization and translation, while *networked society of knowledge*[7] is the form of social structure (nowadays, the term "knowledge economics" has wide spread occurrence). Cybernetics 1.0 *de bene esse* matched the project-technological type of organizational culture, whereas cybernetics 2.0 corresponds to the knowledge-based type (at the new stage of development, organization becomes crucial).

Consider the correlation of the two basic categories in the definition of cybernetics 2.0 ("organization" and "control").

Control is "an element, function of different organized systems (biological, social, technical ones) preserving their definite structure, maintaining activity mode, implementing a program, a goal of activity." Control is "an impact on a controlled system, intended for ensuring its necessary behavior" [157].

---

[7]The author believes that "the knowledge-based type of organizational culture," "knowledge society," "knowledge management" and others are lame terms in this context. Really, a preceding type of organizational culture—the professional (scientific) one—was also founded on scientific knowledge. Nevertheless, these terms are widely used. Let us clarify the meaning of knowledge here. In the professional (scientific) type of organizational culture, the leading role belonged to scientific knowledge in the form of texts. The knowledge-based type of organizational culture operates knowledge of people and organizations about activity organization.

Fig. A.2 Organization and control

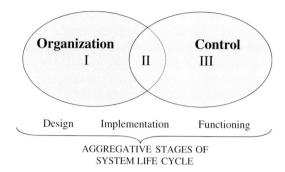

Consequently, the categories of organization and control do intersect, but do not coincide. The former fits system design and the latter fits system functioning[8]; they are jointly realized during system implementation and adaptation, see Fig. A.2. In other words, organization (strategic loop) "foregoes" control (tactical loop).

The domains in Fig. A.2 have the following content (as examples):

I. Design (construction) of systems (including their stuff, structure and functions) —organization but not control (despite that theory of control in organizational systems suggests stuff control and structure control).

II. Joint design of a system and a controlled object. Adaptation. Control mechanisms adjustment.

III. Functioning of controllers in technical systems-control but not organization.

Organization and control can have a "hierarchical" correlation.[9] On the one part, control process calls for organization (organization as a stage in Fayol's management cycle and a function of organizational control, see [131]). On the other part, organization process (e.g., system life cycle) might and should be controlled.

Following the complication of systems created by mankind, the process and property of organization will attract more and more attention. Indeed, control of standard objects (e.g., controller design for technical and/or production systems) gradually becomes a handicraft rather than a science; modern challenges highlight standardization of activity organization technologies, creation of new activity technologies, etc. (*activity systems engineering*).

A fruitful combination of organization and control within cybernetics 2.0 would give a substantiated and efficient answer to the primary question of activity systems engineering: how should control systems for them be constructed? Actually, this is

---

[8]A conditional analogy: organization corresponds to deism (the creator of a system does not interfere in its functioning), while control corresponds to teism (the opposite picture).

[9]Generally speaking, the correlation of organization and control is far from trivial and requires further perception. For instance, in multi-agent systems decentralized control (choosing the laws and rules of autonomous agents interaction) can be treated as organization. Another example is the Bible as a tool of organization [174] (a system of norms making common knowledge and implementing institutional control of a society).

a "reflexive" question related to second-order and even higher-order cybernetics. Mankind has to learn to design and implement control systems for complex systems (high-technology manufacturing, product life cycle, organizations, regions, etc.), similarly to the existing achievements in technical systems engineering.

Cybernetics is important from general educational viewpoint, since it forms the integral modern scientific world outlook.

Cybernetics 2.0. We have defined cybernetics 2.0 as the science of (general regularities in) systems organization and their control.

A close connection between cybernetics and general systems theory and systems analysis, as well as the growing role of technologies (see Figs. 1.9, 4.1 and 4.2) leads to a worthy hypothesis. Cybernetics 2.0 includes *cybernetics* (Wiener's cybernetics and higher-order cybernetics discussed in Sect. 1.2), Cybernetics, and *general systems theory* and *systems analysis* with results in the following forms:

- general laws, regularities and principles studied within metasciences— *Cybernetics* and *Systems analysis*;
- a set of results obtained by sciences-components ("umbrella brands"—*cybernetics* and *systems studies* uniting appropriate sciences);
- design principles of corresponding technologies.

We discuss the latter in detail. A *technology* is a system of conditions, forms, criteria, methods and means of solving a posed problem [148, 149]. Today technologies standardize *craft/skill*[10] and *art*[11] via identification and generalization of best practices; creation of technologies calls for appropriate scientific grounds, see Fig. A.3.

We separate out the following *general technologies*:

- *systems technologies* (general principles; activity organization);
- *informational technologies* (activity support type);
- *organizational technologies* (coordinated joint activity implementation).

Alongside with general technologies, there exist *"sectoral" technologies* of practical activity ("production"); they depend on application domains and possess specifics.

According to this viewpoint, complex study and design of any systems (whether machines, animals or society) within cybernetics 2.0 employs corresponding results obtained by method- and subject-oriented sciences, as well as by general and sectoral technologies—see Fig. A.4.

Keywords for cybernetics 2.0 are *control*, *organization* and *system* (see Fig. A.5).

Similarly to cybernetics in its common sense, cybernetics 2.0 has a *conceptual core* (Cybernetics 2.0 with capital C). At conceptual level, Cybernetics 2.0 is

---

[10]A craft is a personal skill of routine operations based on experience.
[11]Art is a system of techniques and methods in some branch of practical activity; the process of talent usage; an extremely developed creative skill or ability.

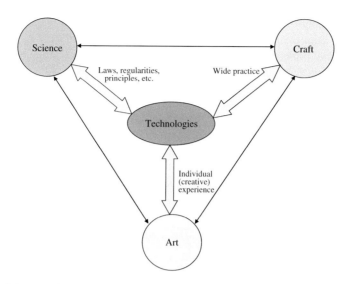

Fig. A.3 Science, technology, craft and art

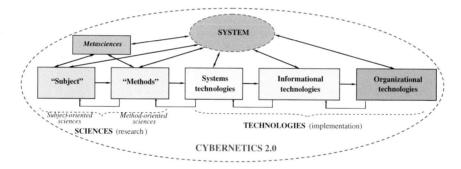

Fig. A.4 Sciences and technologies

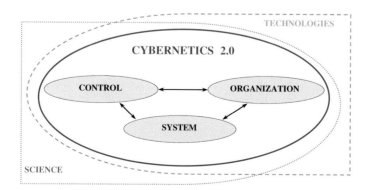

Fig. A.5 Keywords of cybernetics 2.0

composed of control philosophy (including general laws, regularities and principles of control), control methodology, Organization theory (including general laws, regularities and principles of (a) complex systems functioning and (b) development and choice of general technologies), as illustrated by Fig. A.6.

Basic sciences for cybernetics 2.0 are control theory, general systems theory and systems analysis, as well as systems engineering—see Fig. A.6.

Complementary sciences for cybernetics 2.0 are informatics, optimization, operations research and artificial intelligence—see Fig. A.6.

The *general architecture of cybernetics 2.0* (see Fig. A.6) admits projection to different application domains and branches of subject-oriented sciences depending on a class of posed problems (technical, biological, social, etc.).

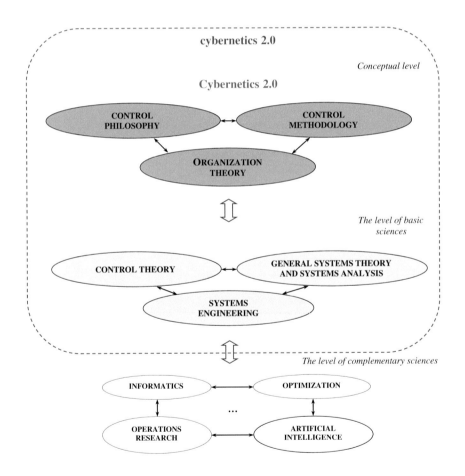

Fig. A.6 The composition and structure of cybernetics 2.0

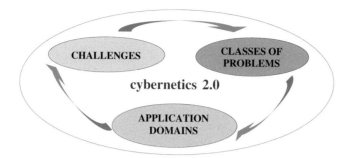

Fig. A.7 The challenges, classes of problems and application domains of cybernetics 2.0

The prospects of cybernetics 2.0. Further development of cybernetics has several alternative scenarios as follows:

- *the negativistic scenario* (the prevailing opinion is that "cybernetics does not exist" and it gradually falls into oblivion);
- *the "umbrella" scenario* (owing to past endeavors, cybernetics is considered as a "mechanistic" (non-emergent) union, and its further development is forecasted using the aggregate of trends displayed by the basic and complementary sciences under the "umbrella brand" of cybernetics);
- *the "philosophical" scenario* (the framework of new results in cybernetics 2.0 includes conceptual considerations only—the development of conceptual level);
- *the subject-oriented* (sectoral) *scenario* (the basic results of cybernetics are obtained at the junction of sectoral applications);
- *the constructive-optimistic* (desired) scenario (the balanced development of the basic, complementary and "conceptual" sciences is the case, accompanied by the *convergence and interdisciplinary translation of their common results*, with subsequent generation of conceptual level generalizations (realization of Wiener's dream "*to understand the region as a whole,*" see the epigraph to this book).

Let us revert to the trends and groups of subjects mentioned in Sect. 1.3. Note that the development of cybernetics 2.0 in the conditions of intensified sciences differentiation provides the following (see Fig. A.7):

- for scientists specialized in cybernetics proper and the representatives of adjacent sciences: the general picture of a wide subject domain (and a common language of its description), the positioning of their results and promotion in new theoretical and applied fields;
- for potential users of applied results (authorities, business structures): (1) confidence in the uniform positions[12] of researchers; (2) more efficient solution of

---

[12]The diversity and inconsistency of opinions and approaches suggested by experts (subordinates) always confuse customers (superiors).

control problems for different objects based on new fundamental results and associated applied results.

Main challenges are control in social and living systems. Several classes of *control* problems seem topical, namely:

- network-centric systems (including military applications, networked and cloud production);
- informational control and cybersafety;
- life cycle control of complex organization-technical systems;
- activity systems engineering.

Among promising *application domains*, we mention living systems, social systems, microsystems, energetics and transport.

There exists a series of global *challenges* to cybernetics 2.0 (i.e., observed phenomena going beyond cybernetics 1.0), see Chap. 5:

1. **the scientific Tower of Babel** (interdisciplinarity, differentiation of sciences; in the first place, in the context of cybernetics—sciences of control and adjacent sciences);
2. **centralization collapse** (decentralization and networkism, including systems of systems, distributed optimization, emergent intelligence, multi-agent systems, and so on);
3. **strategic behavior** (in all manifestations, including interests inconsistency, goal-setting, reflexion and so on);
4. **complexity damnation** (including all aspects of complexity and nonlinearity[13] of modern systems, as well as dimensionality damnation—big data and big control).

Thus, the main *tasks of cybernetics 2.0* are developing the basic and complementary sciences, responding to the stated global challenges, as well as advancing in appropriate application domains, see Fig. A.7.

And here are the main *Tasks of Cybernetics 2.0*:

1. ensuring the Interdisciplinarity of investigations (with respect to the basic and complementary sciences, as illustrated by Fig. A.6);
2. revealing, systematizing and analyzing the general laws, regularities and principles of control for different-nature systems within control philosophy; this would require new and new generalizations (see Fig. 1.10);
3. elaborating and refining Organization theory (O^3).

This book has described the phylogenesis of a new stage of cybernetics–cybernetics 2.0. Further development of cybernetics would call for considerable joint effort of mathematicians, philosophers, experts in control theory, systems engineering and many others involved.

---

[13]Figuratively, in this sense cybernetics 2.0 has to include nonlinear automatic control theory studying nonlinear decentralized objects with nonlinear observers, etc.

Appendix A
A List of Basic Terms[14]

ACTIVITY is an energetic interaction of a human being with an environment, where the former plays the role of a subject exerting a purposeful impact on an object and satisfies its needs. The basic structural components of activity are illustrated by Fig. 2.4.

ADAPTATION is a process establishing or maintaining system's adjustment (i.e., keeping up its key parameters) under changing conditions of an external and internal environment. Quite often, the term "adaptation" means the result of such process-system's fitness to some factor of an environment. The notion of adaptation was pioneered in the context of biological systems, first of all, a separate organism (or its organs and other subsystems) and then a population of organisms. Following the appearance of cybernetics, where an adaptation mechanism is a negative feedback loop ensuring a rational response of a complex hierarchical self-controlled system to varying conditions of an environment, the notion of adaptation has become widespread in social and technical sciences.

ANALYSIS is a mental operation which decomposes a studied whole into parts, separates out particular attributes and qualities of a phenomenon or process, relations of phenomena or processes. Analysis procedures represent an integral component in any study of an object and usually form its first phase: a researcher passes from object exploration as a whole to revelation of its structure, composition, properties and attributes. Analysis is a theoretical method-operation inherent to any activity.

BEHAVIOR is one of several sequences of movements or actions possible in given conditions (a given environment). Behavioral phenomena are inseparably linked with the environment they take place in. Sometimes, human behavior means only the external manifestation of human activity.

BLACK BOX is a system whose internal structure and mechanism of functioning are very complicated, unknown or negligible within the framework of a given problem (i.e., only external behavior makes sense).

CONTROL is (1) an element, function of different organized systems (biological, social, technical ones) preserving their definite structure, maintaining activity mode, implementing a program, a goal of activity; an impact on a controlled

---

[14]Analysis methods for the terminological structure of a subject area were studied in [74].

© Springer International Publishing Switzerland 2016
D.A. Novikov, *Cybernetics*, Studies in Systems, Decision and Control 47,
DOI 10.1007/978-3-319-27397-6

system, intended for ensuring its necessary behavior; (2) the science of control; (3) an object, i.e., a tool of control, a structure (e.g., a department) of several subjects performing control.

DEVELOPMENT is an irreversible, directed and consistent change of material and ideal objects. Development in a desired direction is called progress. Development in an undesired direction is called a regress.

DIVERSITY is a quantitative characteristic of a system, which equals the number of its admissible states or the logarithm of this number.

EXTERNAL ENVIRONMENT is a set of all objects and subjects lying outside a given system, whose behavior and/or changed properties affects the system and all objects/subjects whose behavior and/or properties vary depending on system's behavior.

FEEDBACK (FB) is a reverse impact exerted by the results of a certain process on its behavior; information on the state of a controlled system, which is supplied to a control system (see CONTROL). FB characterizes control systems in wild life, society and technology. There exist positive and negative FB. If the results of a process strengthen its effect, FB is positive. Negative FB takes place whenever the results of a process weaken its effect. Negative FB stabilizes process behavior, whereas positive FB often accelerates process evolution and causes oscillations. In complex systems (e.g., social or biological ones), it seems difficult or even impossible to identify FB types. In addition, FB loops are classified based on the character of bodies and media realizing them: mechanical (e.g., the negative FB realized by Watt's steam engine governor); optical (e.g., the positive FB realized by an optical cavity in a laser); electrical, and others. The notion of FB as a form of interaction plays an important role in the analysis of complex control systems (their functioning and development) in wild life and society.

FUNCTION is (1) (philosophy) a phenomenon dependent on another phenomenon, which varies simultaneously with the latter; (2) (mathematics) a law assigning a certain well-defined quantity to each value of a variable (argument), as well as this quantity itself; a ratio of two (or more) objects such that variation of one object causes an appropriate variation of another object (other objects); (3) a job performed by an organ or organism; (4) a role or meaning of something; a role a subject or a social institute plays with respect to the needs of an upper subsystem or the interests of its groups and individuals; a duty or circle of activity.

GOAL is anything strived for or to-be-implemented. In philosophy, a goal (of an action or activity) is an element in the behavior and conscious activity of a human being, which characterizes anticipation in thinking of the activity result and ways of its implementation using definite forms, methods and means. A goal represents a way of integrating different actions of a human being into a certain sequence or system.

HIERARCHY (from the Greek εραρχία "rule of a high priest") is a structural organization principle of complex multilevel systems, which lies in ordering the interaction between levels of a system (top-bottom), characterizes the mutual correlation and collateral subordination of processes at different levels and ensures its functioning and behavior in whole.

HOMEOSTAT (from the Greek όμοιος "like, resembling" and στάσις "a standing still ") is (1) the capability of an open system for preserving its internal state invariable via coordinated responses for maintaining a dynamic equilibrium; (2) (biological systems) the permanence of characteristics essential for system's vital activity under existing disturbances in an external environment; the state of relative constancy; the relative independence of an internal environment from external conditions [14, 41, 160].

MODEL (in wide sense) is any image, analog (mental or conditional, e.g., a picture, description, scheme, diagram, graph, plan, map, and so on) of a certain object, process or phenomenon (the original of a given model); a model is an auxiliary object chosen or transformed for cognitive goals, which provides new information about the primary object. Model design proper does not guarantee that the resulting model answers its purposes. For normal functioning, a model must meet a series of requirements such as inherence, adequacy and simplicity.

ORGANIZATION: is (1) the internal order, coordinated interaction of more or less differentiated and autonomous parts of a whole, caused by its structure; (2) a set of processes or actions leading to formation or perfection of interconnections between the parts of a whole; (3) an association of people engaged in joint implementation of a certain program or task, using specific procedures and rules, i.e., mechanisms of operation (a mechanism is a system or device determining the order of a certain activity). The last meaning of the term "organization" is the definition of an organizational system. The category of organization is a backbone element of control theory [157].

SELF-ORGANIZATION is a process leading to creation, reproduction or perfection of complex system organization. Self-organization processes run only in systems having a high level of complexity and a large number of elements with nonrigid (e.g., probabilistic) connections. Self-organization properties are inherent to objects of different nature, namely, a living cell, an organism, a biological population, biogeocenosis, a collective of human beings, complex technical systems, etc. Self-organization processes run via readjusting the existing connections and forming new connections among system elements. A distinctive feature of such processes is their purposeful, yet natural (spontaneous) character. Self-organization processes imply system interaction with an external environment, are somewhat autonomous and relatively independent from an environment.

SELF-REGULATION is generally defined as reasonable functioning of living systems; it represents a closed control loop (see FEEDBACK), where the subject and object of control do coincide. Self-regulation has the following structure: an activity goal accepted by the subject, a model of significant activity conditions, a program of actions proper, a system of activity efficiency criteria, information on real results achieved, an assessment of the existing correspondence between real results and efficiency criteria, decisions on the necessity and character of activity corrections.

STRUCTURE is a set of stable connections among the elements of a certain system, ensuring its integrity and self-identity.

SYNERGETICS is an interdisciplinary research direction of self-organization processes in complex systems, which describes and explains the appearance of qualitatively new properties and structures at the macrolevel as the result of interactions among the elements of an open system at the microlevel. Synergetics employs the framework of nonlinear dynamics (including catastrophe theory) and nonequilibrium thermodynamics.

SYNTHESIS is a mental operation which integrates different elements or sides of a certain object in a comprehensive whole (a system). Synthesis appears opposite to and has an indissoluble connection with analysis. Synthesis represents a theoretical method-operation inherent to any activity.

SYSTEM is a set of elements having mutual relations and connections, which forms a definite unity and is dedicated to goal achievement. Systems have the following basic features: integrity, relative isolation from an external environment, connections with the environment, the existence of parts and their connections (structuredness), whole system dedication to goal achievement.

UNCERTAINTY is the absence or incomplete definition or information.

Appendix B
Topics for Further Self-study

(1) The scientific discoveries of the 20th century. The interdisciplinary translation of results
(2) Ampere's cybernetics
(3) Trentowvski's cybernetics
(4) Bogdanov's tectology
(5) N. Wiener and its contribution to cybernetics
(6) W. Ashby and its contribution to cybernetics
(7) S. Beer and its contribution to cybernetics
(8) L. von Bertalanffy and general systems theory
(9) H. Foerster and general systems theory
(10) A. Berg and its contribution to cybernetics
(11) V. Glushkov and its contribution to cybernetics
(12) A. Kolmogorov and its contribution to cybernetics
(13) A.A. Lyapunov and its contribution to cybernetics
(14) The history of controller theory
(15) The history of control theory
(16) The history of general systems theory and systems analysis
(17) The history of informatics
(18) The history of artificial intelligence
(19) The history of operations research
(20) The history of cybernetics in the USSR and USA
(21) The history of systems science and systems engineering
(22) Ontological analysis of basic definitions in cybernetics
(23) Systems of systems
(24) Bibliometric analysis of general cybernetics and applied cybernetics
(25) Bibliometric analysis of conferences on cybernetics
(26) Second-order cybernetics
(27) Autopoiesis
(28) Third- and higher-order cybernetics
(29) Economic cybernetics
(30) Cybernetical physics
(31) Control philosophy
(32) Control methodology

© Springer International Publishing Switzerland 2016
D.A. Novikov, *Cybernetics*, Studies in Systems, Decision and Control 47,
DOI 10.1007/978-3-319-27397-6

(33) The philosophy and methodology of informatics. Information philosophy
(34) The methodology of "soft" systems
(35) Boulding's system classes
(36) Systems dynamics
(37) Laws, regularities and principles of control
(38) Solution methods for weakly formalized problems
(39) Hybrid models. The multimodel approach. Hierarchical modeling
(40) "Hard" and "soft" models
(41) Organization theory
(42) Emergent intelligence
(43) Big data and control problems.

References

1. Ackoff, R.: Towards a system of systems concepts. Manage. Sci. **17**(11), 661–671 (1971)
2. Ackoff, R., Emery, F.: On Purposeful Systems: An Interdisciplinary Analysis of Individual and Social Behavior as a System of Purposeful Events, 2nd edn, p. 303. Aldine Transaction, New York (2005)
3. Albertos, P., Mareels, I.: Feedback and Control for Everyone, 318 p. Springer, Berlin (2010)
4. Nisan, N., Roughgarden, T., Tardos, E., Vazirani, V.: Algorithmic Game Theory, 776 p. Cambridge University Press, New York (2009)
5. Amosov, N.: Modeling of Complex Systems, 81 p. Naukova Dumka, Kiev (1968) (in Russian)
6. Ampère, A.-M.: Essai sur la philosophie des sciences, pp. 140–142. Chez Bachelier, Paris (1843)
7. Anokhin, P.: Anticipatory reflection of reality. Russian Studies in Philosophy. No. 7. pp. 97–112 (1962) (in Russian)
8. Anokhin, P.: The Center-Periphery Problem in the Modern Physiology of Neural Activity. Gorky, pp. 9–70 (1935) (in Russian)
9. Anokhin, P.: Theory of Functional System as a Premise of Physiological Cybernetics Development. Biological Aspects of Cybernetics, pp. 74–91. USSR Academy of Sciences, Moscow (1962) (in Russian)
10. Antomonov, Y.: Modeling of Biological Systems: A Handbook, p. 259. Naukova Dumka, Kiev (1977) (in Russian)
11. Arbib, M.: The Metaphorical Brain: An Introduction to Cybernetics as Artificial Intelligence and Brain Theory, p. 384. Wiley, New York (1972)
12. Arrow, K.: Social Choice and Individual Values, p. 99. Wiley, New York (1951)
13. Asaro, P.: Whatever Happened to Cybernetics. Geist in der Maschine, pp. 39–50. Verlag Turia, Wien (2010)
14. Ashby, W.: An Introduction to Cybernetics, p. 295. Chapman and Hall, London (1956)
15. Ashby, W.: Design for a Brain: The Origin of Adaptive Behavior, p. 298. Wiley, New York (1952)
16. Astrom, K., Murray, R.: Feedback Systems: An Introduction for Scientists and Engineers, p. 408 (2012). Princeton University Press, Princeton. http://press.princeton.edu/titles/8701.html
17. Baker, K., Kropp, D.: Management Science: Introduction to the Use of Decision Models, p. 650. Wiley, New York (1985)
18. Barabanov, I., Korgin, N., Novikov, D., Chkhartishvili, A.: Dynamic models of informational control in social networks. Autom. Remote Control. **71**(11), 2417–2426 (2010)
19. Bar-Yam, Y.: Multiscale variety in complex systems. Complexity **9**(4), 37–45 (2004)
20. Bateson, G.: Steps to an Ecology of Mind, p. 542. Chandler Publication Co., San Francisco
21. Bauer, E.: Theoretical Biology, p. 206. All-USSR Institute of Experimental Medicine, Moscow, Leningrad (1935) (in Russian)

© Springer International Publishing Switzerland 2016
D.A. Novikov, *Cybernetics*, Studies in Systems, Decision and Control 47,
DOI 10.1007/978-3-319-27397-6

22. Beer, S.: Brain of the Firm: A Development in Management Cybernetics, p. 319. Herder and Herder, London (1972)
23. Beer, S.: Cybernetics and Management, p. 214. The English University Press, London (1959)
24. Bernstein, N.: Sketches on the Physiology of Movements and the Physiology of Activity, 347 p. Meditsina, Moscow (1966) (in Russian)
25. Bertalanffy, L.: General system theory—a critical review. Gen. Syst. **7**: 1–20 (1962)
26. Bertalanffy, L.: General System Theory: Foundations, Development, Applications, 296 p. George Braziller, New York (1968)
27. Bertalanffy, L.: The theory of open systems in physics and biology. Science **111**, 23–29
28. Blauberg, I., Yudin, E.G.: The formation and essence of systems approach, 271 p. Nauka, Moscow (1973) (in Russian)
29. Bogdanov, A.: Algemeine Organisationslehre (Tektologie). Hirzel, Berlin (1926). I; 1928. II
30. Boulding, K.: General System Theory—The Skeleton of Science. Manag. Sci. **2**, 197–208 (1956)
31. Boxer P., Kenny V.: Lacan and Maturana: constructivist origins for a 3.0 cybernetics. Commun. Cogn. **25**(1), 73–100 (1992)
32. Boyd, S., Parikh, N., Chu, E., et al.: Distributed optimization and statistical learning via the alternating direction method of multipliers. Found. Trends Mach. Learn. **3**(1), 1–122
33. Breer, V.V., Novikov, D.A., Rogatkin, A.D.: Micro- and macromodels of social networks. Automation and Remote Control. Part 1: General Theory; Part 2: Identification and Simulation Experiments (2015)
34. Breer, V.V., Novikov, D.A., Rogatkin, A.D.: Stochastic models of mob control. Large-Scale Syst. Control. **52**, 85–117 (2014) (in Russian)
35. Bubnicki, Z.: Modern Control Theory, 423 p. Springer, Berlin (2005)
36. Burkov, V.: Foundations of Mathematical Theory of Active Systems, 255 p. Nauka, Moscow (1977) (in Russian)
37. Burkov, V., Enaleev, A.: Stimulation and decision-making in the active systems theory: review of problems and new results. Math. Soc. Sci. **27**, 271–291 (1994)
38. Burkov, V., Goubko, M., Korgin, N., Novikov, D.: Introduction to Theory of Control in Organizations, 352 p. CRC Press, New York (2015)
39. Burkov, V., Lerner, A.: Fairplay in Control of Active Systems/Differential Games and Related Topics, pp. 164–168. North-Holland Publishing Company, Amsterdam, London (1971)
40. Buslenko, N.: Modeling of Complex Systems, 420 p. Nauka, Moscow (1978) (in Russian)
41. Cannon, W.: The Wisdom of the Body, 312 p. Norton, New York (1932)
42. Casti, J.: Connectivity, Complexity and Catastrophe in Large-Scale Systems, 203 p. Wiley, Chichester (1979)
43. Checkland, P.: Soft system methodology: a thirty years retrospective. Syst. Res. Behav. Sci. **17**, 11–58 (2000)
44. Checkland, P.: Systems Thinking, Systems Practice, 331 p. Wiley, Chichester (1981)
45. Chernavsky, D.: Synergetics and Information, 288 p. Editorial URSS, Moscow (2004) (in Russian)
46. Chernyak, Y.: Systems Analysis in Economy Management, 191 p. Ekonomika, Moscow (1975) (in Russian)
47. Daft, R.: Organization Theory and Design, 11th edn, 688 p. Cengage Learning, New York (2012)
48. Dancoff, S., Quastler, H.: The Information Content and Error Rate of Living Things/Essays on the Use of Information Theory in Biology, pp. 263–274. University of Illinois Press, Illinois (1953)
49. Dennis, A., Wixom, B., Roth, R.: Systems Analysis and Design, 5th edn, 594 p. Wiley, New York (2012)
50. Diev, V.: Control. Philosophy. Society. Voprosy Filosofii. **8**, 35–41 (2010) (in Russian)

51. Dorf, R., Bishop, R.: Modern Control Systems. 12th edn, 1111 p. Prentice Hall, Upper Saddle River (2011)
52. Druzhinin, V., Kontorov, D.S.: Introduction to conflict theory, 288 p. Radio i Svyaz', Moscow (1989) (in Russian)
53. Namatame, A., Kurihara, S., Nakashima, H. (eds): Emergent Intelligence of Networked Agents, 261 p. Springer, Berlin (2007)
54. Foerster, H.: The Cybernetics of Cybernetics, 2nd edn, 228 p. Future Systems, Minneapolis (1995)
55. Foerster, H.: Understanding Understanding: Essays on Cybernetics and Cognition, 362 p. Springer, New York (2003)
56. Forrest, J., Novikov, D.: Modern trends in control theory: networks, hierarchies and interdisciplinarity. Adv. Syst. Sci. Appl. 12(3), 1–13 (2012)
57. Forrester, J.: Industrial Dynamics, 464 p. Pegasus Communications, Cambridge (1961)
58. Forrester, J.: Principles of Systems, 387 p. Pegasus Communications, Cambridge (1968)
59. Fradkov, A.: Cybernetical Physics: From Control of Chaos to Quantum Control (Understanding Complex Systems), 236 p. Springer, Berlin (2006)
60. Gelfand, I., Gurfinkel, V.S., Tseitlin, M.L.: On Tactics of Complex Systems Control in Connection with Physiology/Biological Aspects of Cybernetics, pp. 66–73. USSR Academy of Sciences, Moscow (1962) (in Russian)
61. George, F.: The Brain as a Computer, 437 p. Pergamon Press, New York (1962)
62. George, F.: The Foundations of Cybernetics, 286 p. Gordon and Breach Science Publisher, London (1977)
63. George, F.H.: Philosophical Foundations of Cybernetics, 157 p. Abacus Press, Kent (1979)
64. Germeier, Y.: Non-Antagonistic Games, 1976, 331 p. D. Reidel Publishing Company, Dordrecht (1986)
65. Gerovich, S.: From Newspeak to Cyberspeak: A History of Soviet Cybernetics, 383 p. MIT Press, Cambridge (2002)
66. Gershenson, C., Csermely, P., Érdi, P., Knyazeva, H., Laszlo, A.: The past, present and future of cybernetics and systems research// Systems. Connecting Matter Life Culture Technol. 1(3), 4–13 (2013)
67. Gigch, J.: Applied General Systems Theory, 2nd edn, 736 p. Harper & Row, New York (1978)
68. Glushkov, V.: Introduction to Cybernetics. Ukr. SSR Academy of Sciences, Kiev 1964 (in Russian)
69. Gonçalves, C.: Quantum cybernetics and complex quantum systems science—A quantum connectionist exploration. Neuroqantology. 13(1), 35–48 (2015)
70. Goode, H., Machol, K.: System Engineering: An Introduction to the Design of Large-scale Systems, 551 p. McGraw-Hill Book Company, New York (1957)
71. Gorsky, Y.: A System-Informational Analysis of Control Processes, 327 p. Nauka, Novosibirsk (1988) (in Russian)
72. Grössing, G.: Quantum Cybernetics. Toward a Unification of Relativity and Quantum Theory via Circularly Causal Modeling, 153 p. Springer, New York (2000)
73. Gubanov, D., Korgin, N., Novikov, D., Raikov, A.: E-Expertise: Modern Collective Intelligence, 150 p. Springer, Heidelberg (2014)
74. Gubanov, D., Makarenko, A., Novikov, D.: Analysis methods for the terminological structure of a subject area. Autom. Remote Control. 75(12), 2231–2247 (2014)
75. Gubanov, D., Novikov, D., Chkhartishvili, A.G.: Social Networks: Models of Informational Influence, Control and Confrontation, 228 p. Fizmatlit, Moscow (2010) (in Russian)
76. Guide to the Systems Engineering Body of Knowledge (SEBoK) vol 1.3.2, 971 p. BKCASE, INCOSE (2015)
77. Haken, H.: Advanced Synergetics: Instability Hierarchies of Self-Organizing Systems and Devices. 2nd edn, 356 p. Springer, New York (1993)

78. Fishwick, P. (ed.) (2007) Handbook of Dynamic Systems Modeling, 760 p. CRC Press, New York (2007)

79. Kharitonov, V., Alekseev, A.O.: The Concept of Subject-Oriented Control in Social and Economic Systems/Polythematic Electronic Journal of Kuban State Agricultural University [Electronic source]. Kuban State Agricultural University, Krasnodar (2015). No. 05 (109). IDA [article ID]: 1091505043. Available at http://ej.kubagro.ru/2015/05/pdf/43.pdf (in Russian)

80. Heylighen, F.: Principles of systems and cybernetics: an evolutionary perspective. Cybernetics and Systems '92, pp. 3–10. World Science, Singapore (1992)

81. Heylighen, F., Joslyn, C.: Cybernetics and second-order cybernetics. Encyclopedia of Physical Science and Technology, pp. 155–170. 3rd edn. Academic Press, New York (2001)

82. Hillier, F., Lieberman, G.: Introduction to Operations Research, 8th edn, 1061 p. McGraw-Hill, Boston (2005)

83. Gertler, J. (ed.): Historic Control Textbook, 304 p. Elsevier, Oxford (2006)

84. Vus, M.A.: The History of Informatics and Cybernetics in Saint Petersburg (Leningrad). Vol. 1. Striking Historical Examples. Ed. by Corr. Member of RAS R.M. Yusupov; Institute of Informatics and Automation of RAS. St. Petersburg: Nauka, 2008. 356 p. (in Russian)

85. Fet, Y. (ed.): The History of Cybernetics, 339 p. Geo, Novosibirsk, (2006) (in Russian)

86. Hitchins, D.: Putting Systems to Work, 342 p. Wiley, New York (1993)

87. Il'in, V.: The Philosophy and History of Science, 432 p. Lomonosov Moscow State University, Moscow (2005) (in Russian)

88. Haskins, C. (ed.): INCOSE Systems Engineering Handbook Version 3.2.2—A Guide for Life Cycle Processes and Activities, 376 p. INCOSE, San Diego (2012)

89. Jackson, M.: Social and Economic Networks, 520 p. Princeton University Press, Princeton (2010)

90. Jaradat, R., Keating, C.: A histogram analysis for system of systems. Int. J. Syst. Syst. Eng. **5** (3), 193–227 (2014)

91. Julong, D.: Introduction to grey system theory. J. Grey Syst. **1**, 1–24 (1989)

92. Kahn, H., Mann I.: Techniques of systems analysis, 168 p. RAND Corporation, Santa Monica (1956)

93. Kalman, R., Falb, P., Arbib, M.: Topics in Mathematical System Theory. McGraw Hill Book Co. (1969)

94. Kaufman, A.: Introduction to Fuzzy Arithmetic, 384 p. Van Nostrand Reinhold Company, New York (1991)

95. Kenny, V.: There's nothing like the real thing. revisiting the need for a third-order cybernetics. Constructivist Found. **4**(2), 100–111 (2009)

96. Khalil, H.: Nonlinear Systems, 2nd ed, 734 p. Prentice Hall, Upper Saddle River (1996)

97. Klaus, G.: Kybernetic und Gesellschaft, 384 p. Veb Deutscher Verlag der Wissenschaften, Berlin (1964)

98. Klaus, G.: Kybernetik in Philosophischer Sicht, 491 p. Dietz Verlag Berlin, Berlin (1961)

99. Kobrinsky, N., Maiminas, E.Z., Smirnov, A.D.: Economic Cybernetics, 408 p. Ekonomika, Moscow (1982) (in Russian)

100. Kogan, A., Naumov, N.P., Rezhabek, V.G., Chorayan, O.G.: Biological Cybernetics, 384 p. Vysshaya Shkola, Moscow (1972) (in Russian)

101. Kolin, K.: Philosophical Problems of Informatics, 270 p. BINOM, Moscow (2010) (in Russian)

102. Kolin, K.: The Formation of Informatics as a Fundamental Science and a Complex Scientific Problem. Sistemy i Sredstva Informatiki. 2006. Special Issue on Scientific and Methodological Problems of Informatics, pp. 7–58 (in Russian)

103. Kolin, K.: The structure of scientific research on the complex problem of informatics. Sotsial'naya Informatika. Higher Commercial School, Moscow. pp. 19–33 (1990) (in Russian)

104. Kolmogorov, A.: Mathematics—A Science and Profession. Kvant. No. 64. Nauka, Moscow (1988). pp. 43–62 (in Russian)
105. Korepanov, V., Novikov, D.: The diffuse bomb problem. Autom. Remote Control. **74**(5), 863–874 (2013)
106. Korepanov, V., Novikov, D.: Models of strategic behavior in the diffuse bomb problem. Control Sci. **2**, 38–44 (2015) (in Russian)
107. Korepanov, V., Novikov, D.: Reflexive colonel Blotto game. Control Syst. Inf. Technol. **1** (47), 55–62 (2012) (in Russian)
108. Korshunov, Y.: Mathematical Foundations of Cybernetics, 496 p. Energoatomizdat, Moscow, (1987) (in Russian)
109. Kozielecki, J.: Psychological Decision Theory, 424 p. Springer, London (1982)
110. Kozlov, V.: Systems analysis, optimization and decision-making, 176 p. Prospekt, Moscow (2010) (in Russian)
111. Krylov, S.: Neocybernetics: Algorithms, Evolution Mathematics and Future Technologies, 288 p. LKI, Moscow (2008) (in Russian)
112. Kuhn, T.: The Structure of Scientific Revolutions, 264 p. University of Chicago Press, Chicago (1962)
113. Kuzin, L.: The Foundations of Cybernetics. Energiya, Moscow (1979). **1**, 504 p., **2**, 584 (in Russian)
114. Larichev, O.: Systems analysis: problems and prospects. Autom. Remote Control. **36**(2), 241–249 (1975)
115. Lefevbre, V.: Algebra of Conscience, 372 p. Springer, London (2001)
116. Lefevbre, V.: Second-Order Cybernetics in the Soviet Union and Western Countries. Reflexive Processes Control **2**(1), 96–103 (2002) (in Russian)
117. Lefevbre, V.: The Structure of Awareness: Toward a Symbolic Language of Human Reflexion, 199 p. Sage Publications, New York (1977)
118. Lepsky, V.: The Philosophy and Methodology of Control in the Context of Scientific Rationality Development. XII All-Russian Meeting on Control Problems, pp. 7785–7796. Trapeznikov Institute of Control Sciences, Moscow (2014) (in Russian)
119. Lerner, A.: Fundamentals of Cybernetics, 294 p. Springer, Berlin (1972)
120. Malinetsky, G., Potapov, A.B., Podlazov, A.V.: Nonlinear Dynamics: Approaches, Results, Expectations, 3rd edn, 280 p. URSS, Moscow (2011) (in Russian)
121. Mancilla, R.: Introduction to Sociocybernetics (Part 1): Third-Order Cybernetics and a Basic Framework for Society. J. Sociocybernetics **42**(9), 35–56 (2011)
122. Mancilla, R.: Introduction to Sociocybernetics (Part 3): Fourth-Order Cybernetics. J. Sociocybernetics **44**(11), 47–73 (2013)
123. Mansour, Y.: Computational Game Theory, 150 p. Tel Aviv University, Tel Aviv (2003)
124. Maruyama, M.: The Second Cybernetics: Deviation-Amplifying Mutual Causal Processes. Am. Sci. **5**(2), 164–179 (1963)
125. Maturana, H., Varela, F.: Autopoiesis and Cognition, 143 p. D. Reidel Publishing Company, Dordrecht (1980)
126. Maturana, H., Varela, F.: The Tree of Knowledge, 231 p. Shambhala Publications, Boston (1987)
127. Maxwell, J.C.: On Governors. In: Proceedings of the Royal Society of London, vol. 16, pp. 270–283 (1868)
128. Mead, M.: The Cybernetics of Cybernetics. In: von Foerster H. et al. (eds.): Purposive Systems, pp. 1–11. Spartan Books, New York (1968)
129. Meadows, D.: Thinking in Systems, 218 p. Earthscan, London (2009)
130. Meadows, D., Randers, J., Behrens, W.: The Limits to Growth, 205 p. Universe Books, New York (1972)
131. Novikov D. (ed.): Mechanism Design and Management: Mathematical Methods for Smart Organizations, 163 p. Nova Science Publishers, New York (2013)

132. Mesarovic, M., Takahara, Y.: General Systems Theory: Mathematical Foundations (Mathematics in Science and Engineering), 322 p. Elsevier (1975)
133. Mesarović, M., Mako, D., Takahara, Y.: Theory of Hierarchical Multilevel Systems, 294 p. Academic, New York (1970)
134. Milner, B.: Theory of Organization, 2nd edn, 480 p. INFRA-M, Moscow (2000) (in Russian)
135. Mirzoyan, R.: Control as a subject of philosophical analysis. Russ. Stud. Philos. **4**, 35–47 (2010) (in Russian)
136. Moiseev, N.: Mathematical Problems of Systems Analysis, 488 p. Nauka, Moscow (1981) (in Russian)
137. Moiseev, N.: People and Cybernetics, 224 p. Molodaya Gvardiya, Moscow (1984) (in Russian)
138. Morris, W.: Management Science: A Bayesian Introduction, 226 p. Prentice Hall, New York (1968)
139. Morse, P., Kimball, G.: Methods of Operations Research, 258 p. Wiley, New York (1951)
140. Müller, K.: The New Science of Cybernetics: A Primer. J. Systemics Cybern. Inform. **11**(9), 32–46 (2013)
141. Myerson, R.: Game Theory: Analysis of Conflict, 568 p. Harvard University Press, London (1991)
142. NASA Systems Engineering Handbook, 360 p. (2007)
143. Nash, J.: Non-cooperative Games. Ann. Math. **54**, 286–295 (1951)
144. Nature (2008) September 3 (Special Issue)
145. Neumann, J., Morgenstern, O.: Theory of Games and Economic Behavior, 776 p. Princeton University Press, Princeton (1944)
146. Nikanorov, S.: Conceptualization of Subject Domains, 268 p. Kontsept, Moscow (2009) (in Russian)
147. Novick, D.: Program Budgeting, 88 p. Harvard University Press, Cambridge (1965)
148. Novikov, A., Novikov, D.: Methodology, 668 p. Sinteg, Moscow (2007) (in Russian)
149. Novikov, A., Novikov, D.: Research Methodology: From Philosophy of Science to Research Design, 130 p. CRC Press, Amsterdam (2013)
150. Novikov, D.: Analysis of Some Leading Conferences on Control Problems. Autom. Remote Control. **12**, 160–166 (2014) (in Russian)
151. Novikov, D.: Big data and big control. Adv. Syst. Stud. Appl. **15**(1), 21–36 (2015)
152. Novikov, D.: Control Methodology, 76 p. Nova Science Publishers, New York (2013)
153. Novikov, D.: Hierarchical Models of Warfare. Autom. Remote Control. **74**(10), 1733–1752 (2013)
154. Novikov, D.: Mechanisms of Functioning of Multilevel Organizational Systems, 150 p. Control Problems Foundation, Moscow (1999) (in Russian)
155. Novikov, D.: Models of strategic behavior. Autom. Remote Control. **73**(1), 1–19 (2012)
156. Novikov, D.: Regularities of Iterative Learning, 98 p. Trapeznikov Institute of Control Sciences RAS, Moscow (1998) (in Russian)
157. Novikov, D.: Theory of Control in Organizations, 341 p. Nova Science Publishers, New York (2013)
158. Novikov, D., Chkhartishvili, A.: Reflexion and Control: Mathematical Models, 298 p. CRC Press, London (2014)
159. Novikov, D., Rusyaeva, E.: Foundations of control methodology. Adv. Syst. Sci. Appl. **12**(3), 33–52 (2012)
160. Novosel'tsev, V.: Control Theory and Biosystems, 319 p. Nauka, Moscow (1978) (in Russian)
161. Ogata, K.: Modern Control Engineering, 5th edn, 905 p. Prentice Hall, Upper Saddle River (2010)
162. Orlovski, S.: Optimization Models Using Fuzzy Sets and Possibility Theory, 452 p. Springer, Berlin (1987)

163. Optner, S.: Systems Analysis for Business Management, 190 p. Prentice Hall, New York (1960)
164. Pareto, V.: Cours d'Economie Politique, vol. 2, 420 p. (1897)
165. Pawlak, Z.: Rough Sets: Theoretical Aspects of Reasoning about Data. Kluwer Academic Publishing, Dordrecht (1991)
166. Peregudov, F., Tarasenko, F.: Introduction to Systems Analysis, 320 p. Glencoe/McGraw-Hill, OH/Columbus (1993)
167. Pervozvansky, A. A.: Course on Automatic Control Theory. Nauka, Moscow (1986) (in Russian)
168. Peters, B.: Normalizing Soviet Cybernetics. Inf. Culture A J. Hist. Vol. **47**(2), 145–175 (2012)
169. Pickering, A.: The Cybernetic Brain, 537 p. The University of Chicago Press, Chicago (2010)
170. Polonnikov, R., Yusupov, R.M.: Will the 20th century perceive cybernetics. Problemy Upravleniya i Informatiki **6**, 132–152 (2001) (in Russian)
171. Polyak, B., Scherbakov, P.: Robust Stability and Control, 303 p. Nauka, Moscow (2002) (in Russian)
172. Polyak, B., Stepanov, O., Fradkov, A.L.: The 19th IFAC World Congress. Autom. Remote Control. **2**, 150–156 (2015) (in Russian)
173. Pospelov, I.: A Preface to Wiener's books "The Human Use of Human Beings. Cybernetics and Society" and "God and Golem", 248 p. Taideks, Moscow (2003) (in Russian)
174. Prangishvili, I.: Systems Approach and System-wide Regularities, 528 p. SINTEG, Moscow (2000) (in Russian)
175. Prigogine, I., Stengers, I.: Order Out of Chaos, 285 p. Bantam Books, New York (1984)
176. Pushkin, V., Ursul, A.D.: Informatics, Cybernetics, Intelligence: Philosophical Sketches, 341 p. Shtiintsa, Kishinev (1989) (in Russian)
177. Rapoport, A.: General System Theory: Essential Concepts & Applications, 250 p. Abacus Press, Kent (1986)
178. Rashevsky, N.: Outline of a New Mathematical Approach to General Biology. Bull. Math. Biophys. **5**, 33–47, 49–64, 69–73 (1943)
179. Fet Ya.I.: A Reading Book on the History of Informatics. In: Mikhailichenko B.G. (ed.) Institute of Computational Mathematics and Mathematical Geophysics, Siberian Branch of RAS, 559 p. Geo, Novosibirsk (2014) (in Russian)
180. Ren, W., Yongcan, C.: Distributed Coordination of Multi-agent Networks, 307 p. Springer, London (2011)
181. Rosenblueth, A., Wiener, N., Bigelow, J.: Behavior, purpose and teleology. Philos. Sci. **10**, 18–24 (1943)
182. Rukov, A.: Models and Methods of Systems Analysis: Decision-Making and Optimization, 352 p. Moscow Institute of Steel and Alloys, Moscow (2005) (in Russian)
183. Rzevski, G., Skobelev, P.: Managing Complexity, 216 p. WIT Press, London (2014)
184. Sadovsky, V.: Foundations of General System Theory, 280 p. Nauka, Moscow (1978) (in Russian)
185. Satzinger, J., Jackson, R., Burd, S.: Introduction to Systems Analysis and Design, 6th edn, 512 p. Course Technology, Boston (2011)
186. Schedrovitsky, G.: Selected Proceedings, 800 p. Higher School of Culturology, Moscow (1995) (in Russian)
187. Shannon, C.A: Mathematical theory of communication. Bell Syst. Tech. J. **27**, 379–423, 623–656 (1948)
188. Shannon, C., Weaver, W.: The Mathematical Theory of Communication, 144 p. University of Illinois Press, Illinois (1948)
189. Shoham, Y., Leyton-Brown, K.: Multiagent Systems: Algorithmic, Game-Theoretic, and Logical Foundations, 504 p. Cambridge University Press, Cambridge (2008)
190. Smuts, J.: Holism and Evaluation, 368 p. Macmillan, London (1926)

191. Sokolov, B., Yusupov, R.M.: Analysis of Interdisciplinary Interaction between Modern Informatics and Cybernetics: Theoretical and Practical Aspects. In: XII All-Russian Meeting on Control Problems, pp. 8625–8636. Trapeznikov Institute of Control Sciences RAS, Moscow (2014) (in Russian)

192. Sokolov, B., Yusupov, R.M.: Neocybernetics in the modern structure of system knowledge. Robototekhnika i Tekhnicheskaya Kibernetika 2(3), 3–10 (2014) (in Russian)

193. Steinbuch, K.: Automat und Mensch, 392 p. Kybernetische Tatsachen und Hypothesen. Springer, Berlin (1963)

194. Strogats, S.: Nonlinear Dynamics and Chaos: With Applications to Physics, Biology, Chemistry, and Engineering (Studies in Nonlinearity), 512 p. Westview Press, Boulder (2001)

195. Surowiecki, J.: The Wisdom of Crowds: Why the Many Are Smarter Than the Few and How Collective Wisdom Shapes Business, Economies, Societies and Nations, 336 p. Doubleday, New York (2004)

196. Systems Engineering Guide, 710 p. MITRE Corporation, Bedford (2014)

197. Volkova, V.N., Emel'yanov, A.A.: Systems Theory and Systems Analysis in Control of Organizations: A Handbook, 848 p. Finansy i Statistika, Moscow (2006) (in Russian)

198. Taha, H.: Operations Research: An Introduction, 9th ed, 813 p. Prentice Hall, New York (2011)

199. Tesler, G.: New Cybernetics, 404 p. Logos, Kiev (2004) (in Russian)

200. Levine, W.: The Control Handbook, 2nd ed, 786 p. CRC Press, New York (2010)

201. Trentowski, B.: Stosunek Filozofii do Cybernetyki, Czyli Sztuki Rządzenia Narodem, 195 p. Warsawa (1843)

202. Tsetlin, M.: Studies on Automata Theory and Modeling of Biological Systems, 316 p. Nauka, Moscow (1969) (in Russian)

203. Turchin, V.: The Phenomenon of Science, 348 p. Columbia University Press, New York (1977)

204. Uemov, A.: Systems Approach and General Systems Theory, 272 p. Mysl', Moscow (1978) (in Russian)

205. Ugolev, A.: Natural Technologies of Living Systems, 317 p. Nauka, Leningrad (1987) (in Russian)

206. Umpleby, S.: A brief history of cybernetics in the United States. Austrian J. Contemp. Hist. 19(4), 28–40 (2008)

207. Umpleby, S.: The science of cybernetics and the cybernetics of science. Cybern. Syst. 21(1), 109–121 (1990)

208. Ursul, A.: The Nature of Information, 288 p. Politizdat, Moscow (1968) (in Russian)

209. Valachich, J., Jeorge, J., Hoffer, J.: Essentials of Systems Analysis And Design, 5th edn, 445 p. Prentice Hall, Pearson (2012)

210. Varela, F.A.: Calculus for self-reference. Int. J. Gen. Syst. 2, 5–24 (1975)

211. Vassilyev, S., Zherlov, A.K., Fedosov, E.A., Fedunov B.E.: Intelligent Control of Dynamic Systems, 352 p. Fizmatlit, Moscow (2000) (in Russian)

212. Vittikh, V.A.: Evolution of Ideas on Management Processes in the Society: From Cybernetics to Evergetics. Group Decision and Negotiation. http://link.springer.com/article/10.1007/s10726-014-9414-6/fulltext.html. Published online on September 14, 2014

213. Volkova, V.: From the history of systems analysis evolution in Russia, 210 p. St. Petersburg State Technical University, St. Petersburg (2001) (in Russian)

214. Volkova, V., Denisov, A.A.: The Foundations of Systems Theory and Systems Analysis, 2nd edn, 512 p. St. Petersburg State Technical University, St. Petersburg (2001) (in Russian)

215. Voronov, A.: Controllability, Observability, Stability, 339 p. Nauka, Moscow (1979) (in Russian)

216. Vyshnegradsky, I.: On Direct-Action Controllers. Izvestiya St. Petersburg Practical Technological Institute 1, 21–62 (1877) (in Russian)

217. Wagner, H.: Principles of Operations Research, 2nd edn 1039 p. Prentice Hall, NJ Upper Saddle River (1975)
218. Walter, G.: The Living Brain, 255 p. Pelican Books, London (1963)
219. Wasson, C.: System Analysis, Design and Development: Concepts, Principles and Practices, 832 p. Wiley, Hoboken (2006)
220. Weibull, J.: Evolutionary Game Theory, 256 p. The MIT Press, Cambridge (1995)
221. Wiener, N.: Cybernetics: or the Control and Communication in the Animal and the Machine, 194 p. The Technology, Cambridge (1948)
222. Wiener, N.: Ex-Prodigy: My Childhood and Youth, 317 p. The MIT Press, Cambridge (1964)
223. Wiener, N.: God and Golem, Inc.: A Comment on Certain Points where Cybernetics Impinges on Religion, 99 p. The MIT Press, Cambridge (1966)
224. Wiener, N.: I Am Mathematician, 380 p. The MIT Press, Cambridge (1964)
225. Wiener, N.: The Human Use of Human Beings. Cybernetics and Society, 200 p. Houghton Mifflin Company, Boston (1950)
226. Wooldridge, M.: An Introduction to Multi-Agent Systems, 376 p. Wiley, New York (2002)
227. Young, S.: Management: A Systems Approach, 360 p. Scott, Foresman and Company, Glenview (1966)
228. Zadeh, L.: Outline of a New Approach to the Analysis of Complex Systems and Decision Processes. IEEE Trans. Syst., Man, Cybern. **3**(1), 28–44 (1973)

Printed in the United States
By Bookmasters